INDIGENOUS KNOWLEDGE
ON ETHNOBOTANY

The Author

Dr. A.K. Ghosh has been working on senescence and source– sink relationship of different types of plants since 1985. Besides he has been working on pollution and ethnomedicine since 1993. With more than 19 years of teaching experiences in H.S. section, he is at present serving as Headmaster in Saraswati Vidyamandir. Dr. Ghosh has already published 50 original scientific papers in International Journals of repute. He is the author of 10 books, good sportsman and a social worker.

He was awarded **Gopal Chandra Bhattacharya Smriti Puraskar for the year 2004** by Department of Science and Technology, Government of West Bengal.

INDIGENOUS KNOWLEDGE ON ETHNOBOTANY

Dr. ASHIS KUMAR GHOSH
M.Sc., M.Ed., Ph.D, Diploma in Naturopathy;
H.M., Habibpur Saraswati Vidyamandir,
MIDNAPORE, PASCHIM MEDINIPUR-721 101 (W.B.)

2013
DAYA PUBLISHING HOUSE®
A Division of
Astral International Pvt. Ltd
New Delhi - 110 002

Published by : **Daya Publishing House®**
A Division of
Astral International Pvt. Ltd.,
4760-61/23, Ansari Road, Darya Ganj
New Delhi-110 002
Ph. 011-43549197, 23278134
E-mail: info@astralint.com
Website: www.astralint.com

Laser Typesetting : **Classic Computer Services,**
Delhi - 110 035

Printed at : **Thomson Press India Limited**

PRINTED IN INDIA

Dedication

The author dedicates this work to the memory of Late **Abinash Chandra Ghosh** (Great Grandfather, Former Headmaster during 1906-1914 of Rajagram S.B. Raha Institution, P.O. – Rahagram, Bankura which is previously known as Rajagram A.S. School) and to late grandfather Bhutnath Ghosh, starter of Calcutta Tram Company and to my parents.

Dr. K.P. Biswas
M.Sc. Ph.D., D.F.Sc. (Bom.)
F.Z.S. (Cal.), E.F. (W. Germany)
Former Joint Director of Fisheries, Govt. of Orissa
Director of Fisheries A&N., Admn., Govt. of India

Visiting Professor :
W.B. University of Animal & Fisheries Science.

Foreword

Sustainable ecofriendly management, the age old practices of the orient, which was suppressed by the influence of western method of practices is now gaining popularity among the western countries itself. Still today, traditional knowledges continue to be an important source for ecological development throughout the world. People in general are now prefer to use natural products. In this perspective, unwritten and undocated traditional practices has tremendous potentialities.

Practically there is no book on this subject hence Dr. Ashis Ghosh's book on "Indigenous Knowledge on Ethnobotany" I presume, will help in guiding those methods of much needed measure for mankind.

Dr. K.P. Biswas
Former Joint Director of Fisheries,
Govt. of India

Acknowledgements

I am privileged to record my gratitude to Prof. A.K. Biswas (Principal of Ramananda College, Bishnupur), Swaswati Sen (Director of WWF, Kolkata), Amalendu Chakraborty (Director of Nandanik Correspondence College, Midnapore), Asim Chakraborty (D.I. office), Dr. Dipanwita Jana, Sudipta Mondal and Himika Palmal (Business Development Executive of Axis Bank), Narayan Santra (H.M.), N. Boxi of Durgapur Steel plant for their keen interest in this work. My sincere thank is also due to my colleagues and M.C. members of Saraswati Vidyamandir for their encouragement. In the preparation of this book I have received valuable suggestions from the following persons : Dr. Subhasis C. Patra, Debdulal Bhattacharya, Sanchita Sau, Dhirendranath Kar, Durgesh Ranjan Roy, and Joydeep Roy (Prof. of Guru Nanak Institute of Pharmaceutical Science & Technology, Sodepur, Kol – 114).

Author wants to express his thanks to Sarathi Maity (A.I/s) and Soma Ghosh for the patience shown by them during the preparation of the manuscript. Thank is also due to all the staff of D.I. office for their encouragement.

Dr. Ashis Kumar Ghosh

Preface

There are a large number of plants whose medicinal properties are known to the ethnic communities. There is no extensive record on the various uses of plants by the tribal communities and herbalists. In the rural areas, a belief still persists that ailments such as stomach ache, vomiting, diarrhoea, dysentery, fever and child diseases are caused by evil spirits and wrath of displeased God. Most of the spiritual and herbal practitioners are engaged in healing ailments of such kinds. Though herbalism is now a fast growing discipline. There is practically no organised system to record information on the various uses of plants and other materials by the ethnic groups. It is strongly felt that the phytochemical, pharmacognosical and pharmacological evaluation of all these plants and materials should be undertaken without any further delay and the active principle of such medicines and materials should be studied. The gradual anthropogenic effects, civilization, depletion of natural habitats, vegetation and forest in West Rarrh has become another threat to the knowledge of ethnomedicine and traditional practices. The present book is an effort to cater the basic knowledge as well as to bring people closer to nature, stimulation of the senses is also very essential.

Dr. Ashis Kumar Ghosh

Contents

1. Introduction

1.1 General

I have spent a large part of my life studying the traditional knowledge and its application for living beings (both human and veterinary) to maintain the ecofriendly management system. I have realised that despite all the advancement of science, the scope for rejuvenating indigenous knowledge systems remains ahead. The information available from our ethnic community has been put forward primarily to encourage the young naturalists and herbalists. Besides indigenous knowledge provides ecofriendly model for rural development. Generally indigenous food habits have the potential to enhance the quality of life. So, there is need to preserve the traditional food system of different living creatures. Keeping this view in mind, a maiden effort has been made in this book to enlighten the economically poor but knowledge rich ethnic groups such as santals, sabars, bhumij, mahali etc.

Naturalists prefer not to use chemical fertilizers, pesticides, synthetic fibres, plastic, pots, containers, drugs etc., but preferring natural alternatives. So, exhaustive effort has been made as far as possible. Continuity of global warming due to emission of Green House Gases (CFC, CO_2, CO, NH_3, N_2O, NO etc.) by the unscientific activities of man has become a major threat of the present day. Considering all the unlawful activities of human, the present book is an extensive effort to cater the basic knowledge for sustainable development as without natural fundamental elements (air, water, soil etc.) and biodiversity man can not exist.

Classification of Crops :

Crop plants are classified in a number of ways. From agronomic point of view, crop plants are classified as follows :

1. *Cereals* : Sources of Carbohydrate :

Paddy (*Oryza sativa*), wheat (*Triticum* sp), Oat (*Avena sativa*), Barley (*Hordeum vulgare*).

2. *Millets* : Sources of Carbohydrate :

Millets are small grained cereals. It is used as stable food in drier regions of developing countries.

Example : Bajra *(Pennisetum typhoides)*, Jowar *(Andropogon sorghium* Boot/ *Sorghum vulgare* per), Maize (*Zea mays*), common millet (*Penicum miliaceum*), Finger millet *(Elensine coracana)*, little millet *(Panicum miliare)* etc.

3. *Oil seed* : Sources of oils :

Rapeseed and mustard (*Brassica* sp.), sesame *(Sesamum indicum/S. orientale)*, Linseed *(Linum usitatissinum)*, Castor *(Ricinus communis)*, Sunflower *(Helianthus annus)*, safflower *(Carthamus tinctorius)*, Groundnut *(Arachis hypogaea)* etc.

4. *Pulses* : Sources of Protein :

Gram *(Cicer arietinum)*, Pea *(Pisum* sp.*)* Red gram *(Cajanus cajan)*, Lentil *(Lens esculenta/L. culinaris)*, Gram *(Phaseolus aureus)*, Black gram *(Phaseolus mungo)*, Cowpea *(Vigna sinensis)*, Soyabean *(Glycine max)* etc.

5. *Sugar Crops* : Sources of Sugar

Sugarcane *(Saccharum officinarum)*, Sugar beet *(Beta vulgaris)* etc.

6. *Fibre Crops* : Sources of Fibre

Jute *(Chorchorus* sp.), Cotton *(Gossypium* sp.*)* Mesta *(Hibiscus* sp.*)* Sunhemn hemp *(Crotalaria juneca)* etc.

7. *Vegetables* : Sources of Vitamins and Minerals

(a) *Country Vegetables :* Pumpkin *(Cucurbita pepo, C.moschala* Duch*)* Karala *(Momordica charantia)*, Cucumber *(Cucumis sativus)*, Lady's finger *(Hibiscus esculentus/Abelmoschus esculentus* Monch) Indian bean or Deshi Shim *(Dolichos* *lablab),* Ridge gourd *(Luffa acutangula),* Pointed gourd *(Trichosanthes dioia),* Snake gourd *(Trichosanthes anguina),* smooth gourd *(Lunffa cylindrica)* etc.

(b) **English or exotic vegetables :** Potato *(Solanum tuberosum),* Cabbage *(Brassica oleracca* var *capitata),* cauliflower *(Brassica oleracea* var. *botytis),* Kholkhol *(Brassica caulorapa/* B. *oleracea* Var. *caulocarpa),* Tomato *(Lycopersicon esculentum), Lettuce (Lactuca sativa),* Radish *(Raphanus sativus),* Carrot *(Daucus carota),* Turnip *(Brassica rapa),* Beet root *(Beta vulgaris),* Spinach *(Spinacea oleraced)* etc.

8. **Spices and condiments :** Sources of Spices Products

Anise *(Pimppinella anisum),* Cumin *(Cuminum cyminum)* Ginger *(Zingiber officinale Rose),* Coriander *(Coriandrum sativum),* Black pepper *(Piper nigrum),* Chilli *(Capsicum frutescens),* Fenugrees *(Trigonella foenumgraecum),* Onion *(Allium cepa),* Garlic *(Allium sativum),* Cardamon *(Eletaria cardamonum* Maton*),* Nutmeg *(Myristica laurifolia/M. fragrans* Hautt.*)* etc.

9. **Pot herb** *(Source of leafy vegetables)*

Spinach *(Spinacia oleracea),* Bottle gourd *(Lagenaria leucantha* Rusby/*L. siceraria* Stanll. Amaranth *(Amaranthus* sp.*)*

10. **Narcotics:** *(Source of narcotic products)*

Tobacco *(Nicotiana tabacum* L) Indian hemp. or ganja *(Cannabis sativa* L), Poppy *(Papaver somniferum* L), Betel vine *(Piper betel.)*

11. **Root crops :** *(Sources of vegetables and spices) :*

Radish *(Raphanus sativus),* Carrot *(Daucus carota* L), *Beet (Beta vulgaris),* Sweet potato *(Ipomoea batatus)* etc.

12. **Stem or tuber crop :** *(Sources of vegetables and spices)*

Potato *(Solanum tuberosum),* ginger *(Zingiber officinale Rosc),* Turmeric *(Curcuma longa/C. domestica* vol.*)*

13. **Forage crops :** *(Sources of Fodder for Cattle)*

Egyptian clover of Berseem *(Trifolium alexandrium* L), Alfalfa or Lucen *(Medico sativa* L), Guinea grass *(Panicum maximum* Jacq.), Napier grass *(Pennisetum purpureum* Schum), Indian clover *(Melilotus parviflora* desv.), Johnson grass *(Sorghum halepense),* Rhodes grass *(Chloris gayana* Kunth), Sudan grass *(Sorghum sudanense* stapf).

14. Dyes (Sources of dye products) :

Safflower *(Carthamus tinctorius),* Mehedi *(Law-sonia alba Lam/* L. Inermis* L), Indigo *(Indigofera sativa* L). etc.

15. Plantation crop *(Sources of beverage and fruit).*

Areca palm *(Areca catechu* L), Cacao *(Theobroma cacao* L.) Cinchona *(Cinchona officinalis* L), Coffee *(Coffea arabieca L),* Para rubber *(Hevea graziliensis; Hevea brasiliensis* Mullang) Tea *(Camellia thea, Camellia sinensis* O ktze.) etc.

16. Medicinal Plants : *(Sources of Medicine)*

Penicillium *(Penicillium notatum),* Sarpagandha *(Rauwolfia serpentina),* Vasaka *(Adhatoda vasica),* Tulsi *(Oscimum sanctum),* Kalmegh *(Andrographis paniculata),* Cinchona *(Cinchona succirubra),* Nux vomica, Belladona, Ipecac. etc.

17. Fruits Crops : *(Sources of vitamins and minerals)*

Mango *(Mangifera indica),* Jackfruit *(Artocarpus integrifolia* L : *Artocarpus heterophyllus* Lam), Apple *(Pyrus malus* L), Banana *(Musa* sp.), Bullock's heart *(Anona reticulata* L), Cape groseberry *(Physalis peruviana* L), Cashew *(Anacardium occidentale* L), Cherry *(Prunus avium* L), Custard apple *(Annona squamosa* L), Date palm *(Phoenix dactylifera,* L), Fig *(Ficus carica* L), Grape fruit *(Vitis vinifera* L), Guava *(Psidium guajava* L), Jujube *(Zizyphus jujuba Lam),* Lemon *(Citrus limonia* ozbeck, *Citrus limon* Burm f), Lime, Acid and sour *(Citrus aurantifolia* swingle), Litchi *(Litchi chinensis* sonn.) Loquat *(Eriobotrya japonica* Lindl), Mangosteen *(Garcinia mangostena* L), Mulbery *(Mulbery morusals L),* Papaya *(Carica papaya* L), Peach *(Prunus persica* Batsch.), Pear *(Pyrus communis* L), Pine apple *(Ananas sativa* Schult; *Ananas comosus* Merr.), Plum *(Prunus domestica* L), Pomegranate *(Prunica granatum* L), Sapota

(Achras sapota L). Pomelo (citrus *decumana* Murr. *Citrus maxima* Merril, *Citrus grandis* Osbeck), Santra orange, Mandarim orange *(Citrus reticulata* Blanco). Strawberry *Fragaria vesca* L), Sweet orange/mosmib *(Citrus sinensis* Osbeck) etc.

18. **Timber Crop :** *(Sources for fuel and timber for furniture making).*

Sal *(Shorea robusta)*, Segoon or Teak *(Tectona grandis)*, Akashmoni *(Acacia auriculiformis)*, Gamar *(Gmelina arborea)*, Sishuoo *(Dalbergia sisso)*, Sirish *(Albizzia lebbec)*, Mehagoni *(Swietenia macrophylla)* etc.

19. **Green manuring crops :** *(Crops used for green manuring)*

Dhaincha *(Sesbania aculeata)*; Sunhemp *(Crotolaria juncea)*, Cowpea *(Vigna catjang)*, Lentil *(Lathyrus sativus)*, Soybean *(Glycine max)*, Maize *(Zea mays)* etc.

20. **Perfume producing crops :** *(Sources of perfume)*

Sandal (White sandal, *Santalum album,* Red Sandal-*Pterocarpus santalinus,* neem *(Azadirachta indica)*, Mutha *(Cyperus rotundus)*.

21. **Beverage crops :** (Sources of Beverage)

Tea *(Cumelia thea)*, Coffee *(Coffea* sp.*)*, Cocoa *(Theobroma cacao)* etc.

22. **Essential oil producing crops :** (Source of essential oil)

Eucalyptus (Eucalyptus sp.), Lemongrass etc.

1.2 Weed and Weed Control

Introduction

India is an agricultural based country. Agriculture is a main source to meet our basic need. Crop cultivation is an important aspect of agriculture to meet the demand of growing population. Crop cultivation suffers due to attack of pest. Pests are of four types as follows :

(1) Weed, (2) Insect (3) Diseases and (4) Other pests e.g. Nematode, Store grain pest, birds, bat, rat etc.

Damages caused by different pests are as follows (in percentage) :

Weed : 45

Insects : 20

Diseases : 30

Other Pests : 5

Definition of Weed

Weeds are defined in different ways by different scientists as follows :

(*i*)   Weed is a plant growing where it is not desired.

– I.S.A. (1987)

(*ii*) Weed is defined as a herbaceous plant not valued for use of beauty, growing wild and rank and regarded as cumberins the ground or hindering the superior vegetation.

– Onions (1956)

(*iii*) Weed is a plant growing out of place where it is not wanted, either because of its inherent disagreeable or poisonous character or because it is taking the place reserved for something else.

– Encyclopedia Americana (1962)

(*iv*) Weed is a general term for troublesome or otherwise undesirable plant usually introduced, grows without intentional cultivation.

– Herbert (1962)

(*v*)   Weed is used by gardeners in the sense of a plant growing without human encouragement, compound names of lower plants/animals.

– Peter (1967)

(*vi*) Weeds are plants growing in places and at times when men wanted either some other plants to grow or no plants at all.

– NAS (1971), Gupta and Lamba (1978), Rao (1983)

(*vii*) Weeds are obnoxious, unwanted, ugly and persistent plants which are injurious for the crop, interfere with agricultural operation, increasing cost of labour and finally reducing the crop yield.

– Robinson (1973)

Weeds may be of two types :

(*i*) **Absolute Weeds :** These are non-useful, unwanted, persistent and harmful in potentialities. *Argemone mexicana, Cyperus rotundus, Mikania scandens, Phularis minor* etc. are the examples of this weed.

(*ii*) **Relative weeds :** These are also known as Rouges. Rouges are off type plants in the crop fields. They are economical plants but not desirable by the cultivators as this type of plant grows in unwanted place. The Jute plant, an economical crops, is grown in paddy field should be roughed out to get the yield of the Principal crop. Even in an area where Ratna is cultivated, off type variety of Pankaj or IR-50 is found to grow, they should be uprooted to maintain the purity of seed and these unwanted plants are termed as weed.

Classification of Weeds

Weeds are classified in different ways as is done with the economic crops. But the following two methods of classification of weed are considered very useful :

1. **Classification of weeds according to ontogeny (Life cycle)** : Weeds are classification into three broad groups as follows :

(*a*) **Annuals :** The weeds that complete their life cycle within a year or season are called *annual weeds*. They are associated with annual crops and propagated mainly through seeds. Annuals are subdivided according to the season of prevalence:

(*i*) **Kharif seasonal annuals or Kharif weed :** The weeds that grow during the summer and attains maturity with the onset of winter and finally dies are called summer annuals. They complete their life cycles during the warm wetseason (*i.e.* from June to September).

Example : *Echinochloa colonum, Echinochloa crusgalli, Heliotropium indicum, Digera arvensis, Cyanotis axillaris etc.*

(*ii*) **Winter seasonal annuals or Rabi weeds :** The weeds that germinate and grow during winter and die at the onset of the summer are called winter annuals. They complete their life cycle during dry season (*i.e.,* from October to February).

Example : *Chenopodium album, Vicia hirsuta, Avena fatua, Phalaris minor, Asphodelus tenuifolius, Melilotus alba* etc.

(*iii*) **Summer seasonal annuals or Zaid weeds :** The weeds that grow and complete their life cycle during hot dry season (*i.e.* February to June) are called summer annuals.

Examples : *Solanum nigrum, Argenome maxicana, Physalis minima, Portulaca oleracea* etc.

(*b*) **Biennials :** Biennials weed complete their life cycle in two years or in two seasons. They grow vegetatively during first year and starts flowering and sets fruits during second year. So they need two distinct seasons, one for vegetative and another for reproductive growth. Biennial weeds are less troublesome than annual weeds as they are harvested along with the crops before flowering and fruiting.

Examples : *Melilotus alba, Eichorium intybus, Alternanthera achinata* etc.

(c) **Perennials :** Perennial weeds live for more than two years and may live almost indefinitely. These weeds may propagate by seeds and vegetative parts like roots, rhizomes, tubers, bulbs etc. or both. Even if the aerial parts of these weeds die or are damaged, the underground parts may continue to live and starts new growth forming flowering shoots and producing seeds year after year. The perennial weeds may be classified into three groups on the basis of their vegetative reproduction.

(i) **Simple perennials :** These perennials may reproduce solely by seeds. Vegetative reproduction occurs only when roots and stems are mechanically out or damaged. The each cut piece may grow individually forming separate roots and stems.

Examples : *Lantana camara, Ipomoea carnea, Dactyloctenum aegyptium etc.*

(ii) **Bulbous perennials :** This perennial propagates through bulbs, bulblets or tuber as well as by seeds.

Examples : *Cyperus rotundus, Allium vineale, Asphodelus tenuifolius* etc.

(iii) **Creeping perennials :** These perennials may propagate by lateral extension of stem along the soil surface, roots or by seeds. In some creeping perennials new shoots may develop either from roots or rhizomes.

Examples : *Cynodon dactylon, Convolvulus arvensis, Imperata cylindrica, Marsilea quadrifolia, Sorghum halepens, Oxalis corniculata* etc.

2. *Classification of weeds according to the type of habitat :* Weeds are of different types according to the type of habitat :

1. **Terrestrial weeds :** The weeds that grow on crop field are known as terrestrial weeds.

Examples : *Chenopodium album, Cyperus rotundus, Melitotus alba, Solanum nigrum, Cynodon dactylon* etc.

2. **Aquatic weeds** : Aquatic weeds are water inhabitant plants. They are classified into the following three types :

   (*i*) **Floating** : This weed float freely either on the water or in a fixed area. They are or may be free floating (*e.g. Eichhornia crassipes, Fistia stratiotes etc.)* and rooted floating (*e.g. Ludwigia adscendens, Trapa bispinosa* etc.)

   (*ii*) **Submerged** : These weeds are anchored to the bottom of the ditch and grow entirely beneath the surface of water.

   Examples : *Vallisneria spiralis, Hydrilla verticillata, Myriophyllum heterophyllum* etc.

   (*iii*) **Emerged** : This weeds have their roots beneath the surface of water but the leaves and stems peep above the water line.

   Examples : *Typha elephantina, Sagittaria sagittifolia, Nelumbo lutea, Monochoria vaginalis* etc.

Characteristics of Weeds

Weeds have the following characteristics :

(*i*) Weeds are hardy and can grow under adverse edaphic, climatic and biotic condition *e.g. Lantana camera, Eupatorium odoratum* etc.

(*ii*) Weeds are in general prolific with abundant seed production potentialities *e.g. Amaranthus* sp., *Chenopodium album* etc.

(*iii*) Weeds propagate vegetatively by rhizomes (*e.g. Cyperus rotundus*, etc.), bulbs (*e.g. Allium vineale*) stem (*e.g. Cynodon dectylon, Paspalum distichum* etc.) as well as by seeds.

(*iv*) The seeds of many weeds a species are specially designed in such a way that they are easily dispersed by wind

water (*e.g. Amaranthus viridis, Amaranthus spinosus, Cyperus cephalodes* etc.) (*e.g. Blumea lacera, Ageratam conyzoides, Vernonia cinerea* etc.) and by animal or human beings (*e.g. Xanthium strumonium, Andropogon aciculatus, Achyranthes aspera* etc.) Some weeds have explosive mechanism for seed dispersal (*e.g. Ruellia prostrata* etc.)

(*v*) Weed seeds remain dormant and viable for 30 to 40 years *e.g. Chenopodium album, Cyperus iria* etc. Dormancy acts as a safely mechanism against adverse environmental conditions.

(*vi*) The weeds have disaggreable flavour and taste (*e.g. Parthenium hysterophorus, Calotropis gigantea* etc.) or equipt with spines or thorns (*e.g. Solanum xanthocarpum, Solanum khasianum* etc.) which is not accepted by cattle and as a result of which these weeds grow luxuriantly under any situation irrespective of soil reaction.

(*vii*) The seeds of some weeds are very similar in shape and size to the crop seeds and their separation is very difficult *e.g. Argemone mexicana* and mustard (*Brassica Asphodelus* sp. and onion (*Allium cepa*), Avena fatua and oat, barley or wheat etc.

(*viii*) Weeds are euryoecious (wide range of tolerance) compared to crop plants which are more stenoecious in nature. Some weeds are able to survive and multiply under such climatic and soil condition where the crop plants can not survive *e.g. Eupatorium odoratum, Glycosmis pentaphylla, Mikania scandens* etc.

(*viii*) The seeds of weed do not germinate at a time. But they germinate in succession and create problem in crop field, *e.g. Chenopodium album* in Potato field.

Dispersal of Weeds

Dispersal means the spreading of weeds from one place to another place. Dispersal of weeds generally takes place through seeds and vegetative parts.

1. **Dispersal of weeds through seeds :** Seeds are protectors and propagators of their kind. Weed seeds remain viable for a long time and their dispersal mainly takes place through natural and human agencies.

A. *Natural agencies :* The following are some of the devices through which the natural agencies work :

(i) *Through wind :* The weed seeds having parachute like structure called pappess are generally dispersed through wind. The seeds of most plant belonging to the compositae family are dispersed by winds from one place to another to invade new areas. The following weeds are generally dispersed through wind : *Ageratum conyzoides, Blumea lacera, Sonchus arvensis, Calotrapis gigantea, Vernonia cineria* etc.

(ii) *Through water :* The aquatic weeds are generally dispersed through water. The weed seeds are carried through irrigation and drainage channels and invade new areas. The following weeds are generally dispersed through water : *Amaranthus viridis, Amaranthus spinosus, Cyperus cephalotes.* The seed of *C. cephalotes* having corky outgrowth in the body floats in water and thus carried to a long distance.

(iii) *By animals :* Animals, birds etc. have a key role in the dispersal of weed seeds. The weed seeds having hooks (*e.g. Xanthium strumonium*), Scaly bracts (*e.g. Achyranthes aspera*), spines (*Chorkanta*) etc. are dispersed by animals.

(iv) **By explosive fruits** : Fruits of some of the weeds (e.g. *Argemone mexicana, Ruellia prostrata* etc.) burst open as soon as the seed attains maturity and shaken out from the pores of the fruits. This type of explosive mechanism helps in the dispersal of weed seeds.

B. *Human agencies* :

(i) **Sowing of impure seeds** : The sowing of crop seed admixtured with weed seed (*e.g. Phalaris minor* in wheat *Argemone mexicana* in mustard, *Echinochloa colonum* in paddy, *Cichorium intybus* in berseem) is one of the most important means of dispersal of weeds. This is occurred if the cultivator is not alert about the purity of crop seed.

(ii) **Through implements** : The seeds or vegetative parts of weeds is carried from one land to another through the agricultural implements (*e.g.* country plough, M.B. Plough, Tractor, Power tiller, wheel hoe etc.) to cause infestation in new areas. e.g. *Dicanthium annulatum, Chenopodium album* etc.

(iii) **Through F.Y.M., compost etc.** : The seeds of some weeds (*e.g. Chenopodium album, Amaranthus viridis, Amaranthus spinosus* etc.) are dispersed through organic manures. The weed seeds may remain viable in the organic manure and carried to new land where its infestation was not there.

2. Dispersal of Weeds through Vegetative Parts

Most of perennial weeds and few annuals are reproduced and spread vegetatively in addition to seeds propagation. The following weeds spread vegetatively through the following vegetative parts :

(i) Rhizome   : *Soghum halepens*

(ii) Tuber   : *Cyperus rotundus, Cyperus iria etc.*

(iii) Runner   : *Cynodon dactylon*

(iv) Bulb   : *Asphodelus tenuifolius*

(v) Creeper   : *Convolvulus arvensis*

Crop-Weed Association

There are some weeds which prefer to grow in crop field and some of them have a distinct association with some specific crops. Following weeds are almost always associated with the respective crop plants :

(i)   Rice : *Echinochloa colonum, Echinochloa crusgalli, Cynodon dactylon, Marsilea* sp., *Impomoea reptans* etc.

(ii) Wheat : *Cynodon dactylon, Chenopodium album, Phalaris minor, Argemone mexicana, Melilotus* sp., *Leucas aspera, Enagalis arvensis* etc.

(iii) Maize, Jowar Bajra and Groundnut : *Cynodon dactylon, Cyperus rotundus, Echinochloa colonum, Eclipta alba, Eleusine indica, Croton sparsiflorum* etc.

(iv) Barley : *Chenopodium album, Phalaris minor, Avena fatua* etc.

(v)   Potato : *Chenopodium album, Cynodon dactylon, Echinoshloa colonum, Melilotus* sp., *Cyperus rotundus, Argemone mexicana, Anagallis arvensis* etc.

(vi) Jute : *Cyperus rotundus, Cynodondactylon, Echinochloa colonum, Euphorbia hirta, Eleusine indica, Digera arvensis* etc.

1.3 Agroforestry

Introduction

Agroforestry means practice of agriculture and forestry on the same piece of land. Agroforestry is a collective name for land use system and technologies where woody perennials (Shrubs, palms, bamboos, trees etc.) are deliberately used on the same piece of land management units as agricultural crops and / or animals in some form of spartial management or temporal sequences. In Agroforesty system, there are both ecological and economical interactions between different components (Lundergren and Raintree, 1982). In India, the word of Agroforestry is often used as a synonym for the farm forestry. Farm forestry is defined as the practice of forestry in all its aspect on the farm or village lands generally integrated with other farm operations (NCA, 1976). In India, Agroforestry is a sustainable land use system which integrates growing of agricultural crops and forest trees together in the same piece of land for maximum production of food, fodder and wood and other products in such a way that the system is economically and ecologically desirable and acceptable to the local population. Vergera (1985) observes that the Agroforestry is a land use System Combining Agriculture and tree crops of varying longevity (ranging from annual through biennials and perennial plants) arranged either temporarily (*i.e.,* crop rotation) or partially (*i.e.,* inter cropping). At a times, live stocks and fish may be added as components. Under Indian conditions, Agroforestry has two main components as follows: (i) Growing of agricultural crops including fodder. (ii) Forestry *i.e.,* growing of trees and woody perennials.

King (1978) suggests that Agroforestry might be considered to be practiced whenever, trees and agricultural crops are grown in mixture provided that the combined with of the rows of agricultural crops do not exceed height of the forest trees at maturity and provided further that the combined width of rows of forest trees do not exceed the height of tree crops at maturity or at the same selected rotations.

Definition of Agroforestry

(*i*) Agroforestry is sustainable land management system which increases the overall yield of land, combines the production of crops (including tree crops) and forest plants and/or animals simultaneously or sequentially on the same unit of land and applied management practices that are compatible with cultural practices of local population.   *–King and Chandler (1978)*

(*ii*) Agroforestry as a land use unit of land and applies management practices that are compatible with the cultural practices of local population.   *–Nair (1979)*

Components of Agroforestry

Agroforestry has some components as follows (Nair, 1989) :

(*i*) It is deliberate growing of woody perennials on the same unit land as agricultural crops and/or animals either in some form of spartial mixture or in sequence.

(*ii*) There must be significant interaction (positive/negative) between the woody and non-woody components of the system either ecological and/or economical.

(*iii*) This is a production system which tends to harmonise the production of various components and also maximises the production from a given unit of land.

(*iv*) The production and use is sustainable and makes use of modern technologies and traditional local experience and is compatible with the social and cultural life of the local population.

(*v*) It is a long term land management systems and cycles of Agroforestry system is more than one year.

(*vi*) Agroforestry is a more complex form of land management both ecologically and economically than other agricultural or forestry systems.

CLASSIFICATION OF AGROFORESTRY SYSTEMS

Nair (1985) classified Agroforestry system on the basis of structure, function, socio-economic and ecological status as follows :

(1) Structural Classification

(i) *Agrisilviculture :* Agrisilviculture means use of land for concurrent production of agricultural crops and forest crops. Agrisilviculture covers all systems in which land is used to produce both forest trees and agricultural crops either simultaneously or alternately (F.A.O., 1978). In India, Agrisilviculture is used for intercropping of forest plantations with agricultural crops in the initial years until the canopy of forest closes. The Agrisilviculture system aims at producing of enough food grains, timber, fodder, fire wood or other products. Important Agrisilviculture may includes the followings :

(a) Growing of multipurpose trees on farm land.

(b) Homestead plantations and home gardens.

(c) Trees with plantation crops.

(d) Multipurpose trees with horticultural crops.

(e) Alley cropping or hedgerow cropping.

The Agrisilviculture systems are more productive and sustainable than agriculture. Agrisilviculture system also includes growing tree crops such as cocoa, coffee, oil palm, coconut, citrus, rubber, papaya etc. with forest trees. In this system, there is a keen competition between two types of crops for water, nutrients, light and space.

In West Bengal, cultivation of direct seeded paddy is *kharif* crop and green gram as autumn crop gave better performance. *Acacia tortilis* can be grown in combination with the fodder crop like Cowpea cluster bean or pearl millet with remunerative return.

(ii) *Agrisilvi Pasture System :* Agrisilvi pasture system is a combination of Agrisilviculture and Silvipasture system. In this system, the land is managed for the concurrent production of agricultural and forest corps and for grazing by domestic animals. In this system, provision is made for the production of food grains, fodder and wood and for grazing cattle.

The components suitable for mild tropical climate are as follows :

(a) *Field Crops :* Paddy (*Oryza sativa*), Maize (*Zea mays* etc.)

(b) *Tree Species : Eucalyptus* sp., *Erythrina suserosa, Morus leaevigata, Putranjiva roxburgi, Robina pseudoncacia* etc.

(c) *Fodder grass : Cymbopogon citratus, Medicago sativa, Spilanthes, Teosinte, Vetiveria zizanoides* etc.

(*iii*) *Silvipastoral System :* The silvipastoral system means a land management system in which forests are managed for the production of wood as well as rearing of domesticated animals. In this system, grass or grass-legume mixture is grown along with woody perennials simultaneously on the same unit of land. Soil improvement in terms of pH, EC, organic carbon and total nitrogen was observed under improved silvipasture as compared to pasture alone.

Example : *Sesbania grandiflora* or *Sesbania sesbans* is grown with *Cenchrus ciliaris, Setaria aneepts, Demanthus* sp. etc.

(*iv*) *Multipurpose Forest Tree Plantation System :* In this system, forest is managed to yield multiple product (*i.e.* flower, fruit, leaves, barks, roots, gums, medicine etc.) in addition to wood. This system is practised particularly in areas where tribal peoples live as they derive from forest not only wood but also large number of other products for their livelihood.

(2) Functional Classification

(1) *Productive Agroforestry System :* In this system, goods are produced predominantly for meeting the basic needs of the society and measures taken to maximise production of annual and perennial crops. This system may include some system of cultivations as follows :

(a) Shifting Cultivation.

(b) Intercropping of trees.

(c) Home of kitchen gardening.

(d) Planting of trees on the boundaries of the field.

(e) Cultivation of plantation crop.

(f) Planting of plant fuel wood, pulp wood etc.

(ii) Protective Agroforestry System : In this system aims at some measures to be taken as follows :

(a) Amelioration of climate.

(b) Reducing soil erosion by water and wind.

(c) Improving the fertility of soil.

(d) Providing shelter and other benefits.

(e) Providing Protection to land

This system includes some important aspect as follows:

(a) Wind break and shelter belt.

(b) Some hedgerow intercropping.

(c) Plantation for soil conservation

(d) Introduction of cover crop.

(iii) Multipurpose Agroforestry System : This system aims at optimising both of the above functions as follows :

(a) Product required for meeting the basic needs such as food, fodder, fuel etc.

(b) Influences for conservation of environment.

Several other systems may be grouped into this class as follows :

(a) Home garden or kitchen garden.

(b) Hedgerow intercropping

(c) Multipurpose tea plantation

(d)   Planting trees on agricultural field

(e)   Silvipasture and several other system.

(3) Socio-economic Classification

(i) *Subsistence Agroforestry System* : This system aims at meeting the basic needs of farmer having small holding and managing different activities by the members of the family. This system includes some system of cultivation as follows :

(a)   Shifting cultivation

(b)   Home gardens

(c)   Homestead forestry

(d)   Planting scattered trees on the field.

(ii) *Commercial Agroforestry System* : In this system, emphasis is given at production of commodities in commercial basis. This system includes plantation crop in combination with agricultural crops.

(iii) *Intermediate Agroforestry System* : This is a system of Agroforestry which is intermediate between subsistence and Commercial Agroforestry. This system includes cultivation of fruit trees with agricultural crops.

Advantages of Agroforestry

Agroforestry System have advantages as follows :

(1) Agroforestry aids in supplementing food and fodder. Mahua tree *(Madhuca indica)* is most important tree in Agroforestry. This tree yields an edible flower, fruit and seeds. The flowers are rich in sugar (65.75%), Calcium (14%), Vitamin B Complex and Vitamin-C (Anon, 1962). A single tree of Mahua can yield about 50-52 kg of dried mahua flower (Anon, 1962). The tribals and other rural people eat the flower of mahua traditionally. Trees of several species are known to yield palatable green leaf fodder (Sen *et al.* 1978), Gulati *et al.* 1982, Singh *et al.*, 1982) as follows :

   (i)   *Acacia nilotica*

  (ii)   *Aegle marmelos*

 (iii)   *Ailanthus excela*

 (iv)   *Bauhina variegata*

  (v)   *Celtis australis*

 (vi)   *Dalbergia sisso*

 (vii)   *Dendrocalamus strictus*

(viii)   *Grewia optiva*

 (ix)   *Handwickia bibata*

  (x)   *Loucama leucocephala*

 (xi)   *Moringa oleifera*

 (xii)   *Morus alba*

(xiii)   *Zizyphus* sp.

Table 1 : Fodder yields from some trees (Singh, 1982)

Names of trees	Yield of fodder per tree per annum (in kilogram)
(i) *Acacia nilotica*	50 – 120
(ii) *Ailanthus excelsa*	100 – 200
(iii) *Bauhinia variegata*	30 – 50
(iv) *Grewia optiva*	25 – 50
(v) *Prosopis cineraria*	25 – 50

Table 2 : Nutritive value of some fodder trees (Sen *et al.*, 1978, Gulati et al. 1982, Singh, 1982)

Sl. No	Tree leaves	Crude protein	Diges- tible crude protein	Total Diges- tible Nutrients	Crude Fibre	Calcium	Phos- phorus
		%	%	%	%	%	%
1	2	3	4	5	6	7	8
1.	Khair *Acacia catechu*	12.0–18.7	2.9	46.3	21.9–22.6	1.6–2.7	0.1–0.2
2.	Babul *Acacia nilotica*	7.0–15.0	–	–	20.1–32.3	1.2–2.6	0.1–0.2
3.	Israeli babul *Acacia tortills*	18.1	–	–	–	3.1	0.2
4.	Haldu *Adina cordifolia*	8.7–15.3	2.0–2.8	50.9	12.1–14.0	1.7–3.2	0.1–0.5
5.	Bel *Aegle marmelos*	15.1–15.3	10.8	56.7	16.5–18.1	4.2–4.8	0.1–0.3
6.	Mahrukh *Ailanthus excelsa*	16.3–19.9	–	–	12.8–21.9	1.5–2.4	0.2–0.3
7.	Ohi *Albizia chinensis*	15.1	4.9	40.2	31.6	1.2	0.1

THINGS TO REMEMBER

We are now facing enormous challenges. The number as well as population size of the planet's 1,313 species of wild vertebrates (reptiles, birds, mammals etc.) is diminishing, the amount of man made chemical substances is increasing and as a result the climate is changing. As many as 37 football ground of tropical rainforest disappear in every minute, 45000 dams have a negative impact on nature, more than 1 billion people do not have access to drinking water and the list is endless. If we continue with the same unsustainable consumption and production pattern, by 2050 we would need 2 planets.

Humanity's ecological footprint is increasing. Our ecological footprint is too big. Since 1980s we consume the world's natural resources at a rate that is 25 per cent faster than nature's capacity to create new ones. Today we produce more waste than nature can deal with, which means that we outstrip nature's capacity to create new resources.

Everyday we impart our environment in different ways, for *e.g.*, when we buy food, drive cars or heat our homes. We affect our surroundings and leave a so-called ecological footprint.

The average North American ecological footprint is 9.4.

The average European ecological footprint is 4.8.

The average Asian ecological footprint is 1.3.

The average African ecological footprint is 1.1.

The average World ecological footprint is 1.8 hectares per person.

How one can reduce his ecological footprint

A lot of things can be done through reducing product consumption, reducing car travel, eating more vegetarian food changing to sustainable energy systems, buying eco-labelled products, supporting environmental organisations, buying second hand clothes etc.

Western life style puts a tremendous burden on the environment. Except a few country life style the world's present development is unsustainable. If we are successfully address the world's problem and possibilities we need to equip ourselves to act for a more sustainable future and to live in harmony with nature. Sustainable development does not only mean ecological sustainability but also includes social and economic dimensions. Sustainable development meets the needs at the present without compromising the ability of future generations to meet their own needs. Besides sustainable development can be regarded as a journey *i.e.*, an ongoing process within an ecological framework. The long-term objective is to have a good life as possible without hampering other living beings, nature and society in both time and space. An economy that is socially unjust or that does not correspond to the ecological framework is not sustainable. In a nutshell, acting sustainable makes economic sense. An economic development that means economic benefits for society as a whole and that does not pose a threat to artificial and natural capital.

There are so many definitions and steps of sustainable development. Such as : (a) care for ourselves (b) care for others (c) care for planet (d) care for offspring and so on. In a global perspective every person seems tiny. But lots of tiny people can make an incredible difference.

Bees and wasps know that the hexagon is the strongest and most flexible shape. They are genetically programmed to construct one cell after another in hexagonal patterns in order to provide the best possible conditions for future generations. In every moment through informal education we learned so more from the natural events. Future generations must be able to cope with the changes through incorporation as education for sustainable development. As learning is an ongoing process and life long so acquiring knowledge might be a demanding and tedious process but it is nevertheless easy to carry - it fits nicely into the body and is personal. Keeping in view "Every sunrise ought to be seen as a new opportunity, every day as a new challenge to secure our planet and its life forms for future generations. Despite all the alarms and investments in intensive environmental work our Earth is still deteriorating.

> *"The earth has enough for man's need*
> *but not for man's greed".*
>
> – Mahatma Gandhi
>
> *"It would be good to live without*
> *spoiling things for others".*
>
> – Wolfgang Brunner

Organic farming for sustainable management

Diversity of plant life is the key factor on organic farms. Besides diversity is essential for soil health. Generally the ground-cover provides shade on sunny days, while leaf litter cools the surface of the soil. High humidity under the canopy of mature long-life trees reduces evaporation and minimize the need for irrigation. The drooping leaves of trees act as a water meter to indicate falling moisture levels.

By ruining the natural soil fertility we actually create expensive artificial needs. So, now-a-days many farmers who are turning back on the increasingly industrialised means of agricultural production that need costly input and taken to organic farming. In the way entire web of life must be protected and nurtured. While organic farming is supposed to be eco-friendly, some one have expressed fears over its possible adverse ecological effects. "We cannot feed 6 billion people with organic farming; if we tried to do so we will level most of our forests" said Norman E. Borlaug, father of the Green Revolution and Nobel Laureate. John Emslay, a British chemist said that the human race will face this century is not global warming but a global conversion to organic farming which would result in the perishing of over 2 billion people. Though India has enough food to feed her population of 1 billion plus, yet hunger and food insecurity at household level have increased. Now farmers unable to pay back debts incurred by the purchase of seeds, pesticides, chemical fertilizers and equipments, kill themselves generally majority of them have committed suicide because they are steeped in debt, from loans taken to sow improved seeds or use better fertilizer or pesticide, or most likely a combination of all the three.

So, organic farming can prove profitable if we return to traditional practices. Now it is most necessary to monitor the health of the soil and thereby increased the beneficial microbial activity. Just I can say "If you save the soil, you save the nutrient. Since many of the developed and developing countries are land-starved, they should immediately switch over to organic farming. The arguments arises that organic farming requires more land. Holds good only for cash crops according to FAO.

In organic farming legumes are grown for N_2 – fixing, and intercropping, crop rotation, composting, vermiculture, and so on, are practised to retain moisture and nutrients. Food productions of today are heavily subsidised. Organic food, since it does not receive any of these subsidies, in comparison, comes across as being found expensive. If the same subsidies are given like that of nonorganically grown foods, and is perhaps likely to be cheaper in view of its superior yield. As a consequence widespread adoption of organic farming is unlikely to materially impact the availability of food. To begin with, the practice of organic farming should be for low volume high value crops like medicinal plants, fruits, vegetables and spices. As a sustainable means of agricultural production and to promote health instead of spreading toxins through chemical fertilizers and pesticides Indian farmers of remote village are still engaged with their traditional practice of organic farming.

Green Chemistry a Healthy Way

Several environmental laws have been passed to protect the environment by controlling our exposure to hazardous substance. Instead of limiting risk by controlling our exposure to hazardous chemicals green chemistry attempts to reduce and preferentially eliminate the hazard thus negating the necessity to control Exposure. It we do not use or produce hazardous substances then the Risk is nil. The fundamental idea of green chemistry is that the designer of a chemical is responsible for considering what will happen to the world after the agent is put in place. Green chemistry has gained a strong footprint in both industry and academia. The design of environmentally gracious products and processes are guided by 12 fundamental approaches of Green Chemistry to achieve the goals. Many of them are mentioned below :–

1. **Prevention :** It is better to prevent waste than to treat the waste after it is exposed.

2. Less Hazardous Chemical synthesis that possess mild or no toxicity to human health and the environment.

3. Designing Eco-friendly chemicals.

4. Designing Safer Solvents and Auxiliaries (separation agent etc.).

5. Use of Renewable Feedstocks rather than developing.

6. Reduce Unnecessary Derivatives whenever possible.

7. Design for Energy Efficiency *i.e.*, Energy requirement should be minimized.

8. Design for Degradation of the chemical products as a result no harmful end products can persist in the environment.

9. Inherently Safer Chemistry for Accident prevention to minimize the releases, fires and explosion.

As Green Chemistry represents our sustainable future so the coming generation of scientists and research scholars need to be trained for eco-friendly techniques. Besides important steps to be taken for Green Chemistry within the school and college curriculum through the following approaches :–

(a) Incentives to students for working on the projects of green chemistry.

(b) Organizing workshops and training courses.

(c) Introduction of the basic concepts of chemical toxicology and the basis of hazards.

(d) Incorporation of green chemistry topics within the professional examinations etc.

The future challenge for chemists as well as for the scientists is to develop products, processes etc. in a sustainable way to improve quality of life of living beings in accordance with the Ecological rules.

2. Environmental Awareness

2.1 Economic evaluation of Makhna (*Euryale ferox* Salish) and Parapata (*Asparagus* sps.) cultivation respectively in North Bengal and South Bengal

Abstract

Makhna (*Euryale ferox* Salish), a rooted hydrophyte has multiple uses in the districts of Malda, Dinajpur, etc., it is nutritious and have medicinal value. A sum of Rs. 25,000/- may be generated per acre by selling its various parts. Whereas a semiheliophyte named Parapata (*Asparagus* sps) is a common cultivated ornamental plant of the family Liliaceae. A sum of Rs. 18,000/- may be fetched per acre in single harvesting by selling its valuable leaves in winter season.

Keywords : Makhna, Wetlands, Ornamental, Parapata, Mesophyte.

Introduction

Kanta-padma or Makhna or Fox-nut (*Euryale ferox* Salish, Nymphaeaceae) is indigenous to the some districts of North Bengal. It is a large perennial spiny erect aquatic herb with big elliptical or orbicular floating leaves. The root-stock is stout, creeping and fixed in the soil. The flowers are generally purple or reddish. Fruits are rounded and spiny. 30-50 black rounded seeds are found in torus. Seeds, pulp is eaten followed by roasted and dehusking. Dry seed dust is the substitute of arrowroot of commerce.[1-3]

Whereas Parapata (*Asparagus* sps) is a indigenous Mesophyte of the Moyna block of Purba Medinipur and the leaves are harvested 5 times in a year. Harvested leaves are being exported to Gujarat, Mumbai, Delhi, Ahmedabad, Chennai,

Jamshedpur, Bhubaneswar etc. The plants are raised well in shady places of upland areas from the collected seedlings of Contai. Parapata can not enduring water lodged condition.

Materials and Methods

Makhna, the rooted hydrophytes are commonly grown through seeds in the wetlands of Harishchandrapur block of Malda both at a depth and distance of 1 meter favourably during the month of December and harvested in the month of May-June. Though the plant is grows automatically in the subsequent years and hence the farmers do not pay extra attention for its cultivation. Generally a large number of seedlings come out from the mature seeds buried inside which can be collected as planting materials.

Results and Discussion

At present there is a great demand of Makhna at Arab and Afganisthan. Besides it is exported from Malda to Delhi, U.P., M.P., Rajasthan as it is a delicious food both for child and old man for its easily digestible property. It is rich in carbohydrate, protein, fat and minerals. By roasting it is used as a parched-rich along with edible salt and pepper. So the proper utilization of the wetlands can be recovered by the cultivation of Makhna. Chemical fertilizers and pesticides used in the different crop cultivation will definitely affect the quality of standing water bodies of Global wetlands thereby reducing its yield.

Cultivation along with harvesting in a scientific way can definitely enhance the productivity. For the restoration of wetlands their resources and biodiversity, a joint stream management and effective research oriented programme with the participation of tribal communities is necessary. Same is true for *Asparagus* sps. As the decorating purpose of Parapata have been well established globally, its conservation and commercial cultivation in West Rarrh is essential. In conclusion it can be said that both the plants promises for economic development of local communities promoting thereby green environment. Streams without green leaves, usually straggling, leaves replaced by linear or acicular cladodes with minute scales are the significant character of Parapata which attracts the inviter in any functions as an auspicious thing.

References

1. Sing H.B., Singh R.S. and Sandhu J.S., Herbal Medicine of Monipur, A colour Encyclopaedia, (Daya Publishing House, New Delhi), 2003.

2. Jain A, Singh R.S., and Singh H.B., Economic evaluation of lotus cultivation in Sanapat Lake, Monipur Valley, Natural Product Radiance, 2004, **3** (6), 418-421.

3. Sinha R. and Lakra V. Edible weeds of tribals of Jharkhand, Orissa and West Bengal, IJTK, 2007, **6** (1), 217-222.

2.2 Selected elephant repellent plants from West Rarrh in agro-forestry

Abstract

The efficacy of some plants against a herd of elephants is worth emulating as they can protect the crops as a natural barrier or hinder.

Keywords: West Rarrh, Natural barrier, Elephant repellent.

Introduction

Interest in ethno botany by an organized manner can overcome the frequent raid in the West Rarrh by the elephant troop coming from the Dalma hill of Jharkhand. As at present elephants have no safety zone to survive due to onset of industrialization, increase in population and pollution along with the gradual depletion of forest and food. So, to save crops along with the locality search for a eco-friendly way is essential. Keeping in view some plants were identified those who are able to combat with the elephant regiment as a natural barrier.

The information's furnished here are based on work carried out during the ethnobotanical field explorations and the observations reported here are original. The botanical identity of plants was authenticated by Prof. M.N. Sanyal (an eminent taxonomist) and the voucher specimens were preserved for future use. From ancient time aboriginals judiciously used these plants.

Results and Discussion

Elephants are very clever and quiet in nature. They are habituated to live in the deep forest. From the ancient time elephants are the indispensable part of human beings and are being used in various domestic purposes and also in the battle field. But with the advent of civilization their necessity gradually declined to the modern people. Elephant drivers are well acquainted with their likes and dislikes. Some elephant-philic plants are the paddy, sugarcane, banana, bamboo, beverage yielding plants etc.

Similarly some elephant-phobic plants are the nettle, ginger, red pepper etc. Generally all the living creatures have a tendency to live within their motherland by natural instinct. As the spacious corridors of the elephants are being gradually replaced and occupied by us so due to lack of free grazing zone they occasionally attack the locality in search (*i.e,* to meet their appropriate conditions) of food and drinks. To prevent as well as to bounce them elephant phobic plants can be raised as a fence in the transition zone of the forest and crop land. Besides elephant-phobic crops should be cultivated in the agricultural field to compensate the socio-economic loss caused by the elephants. The faeces of tiger and lion should be occasionally scattered in their corridor as a targeted object to prevent them. Nevertheless hundred of years ago there was a general fond for elephants' foot impression in the ancestral abode as an auspicious object. But today the superstition is totally changed due to our struggle for existence. Henceforth by the above mentioned plants the barricade should be executed respectively by the forest department and agriculture department.

References

1. Thothathri K, Sen R, Pal DC and Molla A, Selected poisonous plants from the tribal areas of India, BSI, 1985.

2. Viswanathan N and Joshi BS, Toxic constituents of some Indian plants. Current Science, 1983, **52** (1) : 1-8.

Table 2.1 : List of elephant repellent plants

Sl No.	Botancial Name	Family	Local Name	Occurrence
1.	*Mucuna pruriens* L.	Fabaceae	Alkusi	Found in bushes.
2.	*Alangium salviifolium* L.	Alangiaceae	Ankar	Scrub jungle.
3.	*Flacourtia indica* Burm, f.	Flacourtiaceae	Bainchi	Grows on bushes.
4.	*Curcuma aromatica* Salisb.	Zingiberaceae	Ban-halud	Wild in the shady places.
5.	*Curcuma longa* L.	Do	Halud	Cultivated.
6.	*Typhonium trilobatum* L.	Araceae	Ghet-kachu	Common in shady places.
7.	*Colocasia esculenta* L.	Do	Kachu	Cultivated and wild.
8.	*Amorphophallus campanulatus* Roxb.	Do	Ol	Cultivated and wild.
9.	*Zingiber officinale* Rosc	Zingiberaceae	Ada	Cultivated.
10.	*Z. montanum Koenig.*	Do	Ban-ada	Wild
11.	*Z. capitatum* Roxb.	Do	Jangli-ada	Wild
12.	*Agave wightii* Drumm.	Agavaceae	Konga	Grows on laterite soil.
13.	*Capsicum annum* L.	Solanaceae	Lanka	Cultivated.
14.	*Alocasia indica* Roxb.	Araceae	Man-kachu	Wild and cultivated.
15.	*Meyna spinosa* Roxb.	Rubiaceae	Moinakanta	Wild
16.	*Trapa natans* L.	Trapaceae	Paniphal	Aquatic
17.	*Opuntia stricta* Haw.	Cactaceae	Phanimansha	Waste lands.
18.	*Argemone mexicana* L.	Papaveraceae	Shialkanta	Waste lands.

2.3 Butterfly-philic plants in agro-forestry of West Bengal

Abstract

Present communication brings to knowledge about butterfly philic plants of West Rarrh along with the conservation of 30 species of butterflies before being abolishing.

Introduction

Butterflies are the most beautiful creatures of nature for their beautiful wings and they are acceptable to 'Hindus Wedding Ceremony' from the ancient time. They also help to pollinate the flowers by their natural instinct. But they never pollinate all the flowers. Besides they have some species specific choice to lay their eggs on some selected plants. The inhabitants of the study area have rich heritage particularly related to plant utility and the literature survey shows that the region was almost untapped from this point of view.

Study area

The West Rarrh of West Bengal mainly constituted the districts of Bankura, Birbhum, Burdwan, Purba and Paschim Medinipur, Murshidabad and Purulia. It is the extended part of Chhotanagpur plateau. Here the forests lie scattered in small patches between latitudes 21°75' and 24°33' N and longitudes from 85°70' to 87°80' E. Its area is 27,500 km². Here the hills are relic type.

Basically there is a dominance of hot and humid climate along with a short duration of winter (Dec. – Jan.) which is suitable for butterflies. Though some tribes of West Rarrh used cocoons of butterflies as their food.

Materials and Method

During ethnobotanical field explorations (2000 – 2002) I have collected this butterflies along with their host range specific plants and preserved them for future use. Plant specimens along with the butterflies were authenticated by a reputed taxonomist (Dr. M.N Sanyal of Bishnupur, RN College) which is given.

Results and Discussion

The data has been accrued from the study area shows some remarkable features. That 30 species of butterflies are still existing in the West Rarrh. Many of them are in the verge of extinct due to pollution by heavy use of insecticides, deforestation, unconsciousness of human beings etc. Due to rapid increase in human and consequent increased population, biotic interference some species of butterflies are dwindling from their natural habitats. It is therefore, imperative that butterflies which are still in vogue should be documented for obvious reasons. So far my knowledge goes in India there was 1503 species of butterfly. Of them in West Bengal there was 515 species.

Butterflies renowned for their ornamented wings. They never like all the plants for taking nectar and laying the eggs. They generally select native herbs and shrubs having sufficient food for their larva. For the conservation of butterflies and to maintain the diversity of butterflies in the garden butterfly-keepers can selected and cultivated the plants according to the needs of larva. To achieve particular species of butterfly we must have concentrated to some particular species of plants. Some plants suitable for this purpose have been enumerated for 30 species of butterflies.

List of Butterfly-philic Plants

1. *Broyophyllum calycinum* Salisb, Crassulaceae, Pathar-kuchi – Succulent herb having reddish-green flower from March to May.

2. *Calotropis gigantea* L., Asclepiadaceae, Akanda Shrub having purplish – white flowers from March to November.

3. *Carthamus tinctorius* L. , Asteraceae, Kusum – Under shrub having yellow or red flowers from February to April.

4. *Citrus aurantifolia* Swing, Rutaceae, Patilebu – Shrub having white flowers almost throughout the year.

5. *Clerodendrum infortunatum* L., Verbenaceae, Ghentu – Shrub having white tinged with purple flowers from February to May.

6. *Clerodendrum viscosum* Vent, Verbenaceae, Bhant – Shrub having white fragrant flowers from February to May.

7. *Crotalaria retusa* L., Fabaceae, Atasi - Under-shrub having bright yellow flowers from October to January.

8. *Evolvulus numularius* L., Convolvulaceae, Bhuin-ankra – Prostrate shrub having white flowers throughout the year.

9. *Flacourtia indica* Burm. f., Flacourtiaceae, Bainchi - shrub having greenish white flowers from January to March.

10 *Glycosmis arborea* Roxb., Rutaceae, Ash-sheora - Shrub having white fragrant flowers from September to December.

11. *Heliotropium indicum* L., Boraginaceae, Hatisur – Herb having purplish – white flowers from April to January.

12. *Hibiscus rosa-sinensis* L., Malvaceae, Jaba – Shrub having red flowers almost throughout the year.

13. *Impatiens balsamina* L., Balsaminaceae, Dopati – succulent herb having rose coloured of white flowers from August to January.

14. *Ixora coccinea* L., Rubiaceae, Rangan – Shrub having reddish pink flowers throughout the year.

15. *Lantana camara* L., Verbenaceae, Putush – Straggling Shrub having purple flowers almost around the year.

16. *Nerium indicum* Mill, Apocynaceae, Karabi – Shrub having rose or white coloured flowers almost through-out the year.

17. *Ocimum sanctum* L., Lamiaceae, Krishna tulsi – Herb having puplish – white flowers from August to January.

18. *Ricinus communis* L. Euphorbiaceae, Rerhi – Under-shrub having greenish – white flowers from January to February.

19. *Solanum nigrum* L., Solanaceae, Kakmachi or gudkamai– Herb having white flowers almost throughout the year, but mainly in February to July.

20. *Tagetes erecta* L., Asteraceae, Genda – Herb having yellow flowers from November to February.

21. *Tagetes patula* L., Asteraceae, Genda – Herb having red flowers from November to February.

22. *Vitex negundo* L., Verbenaceae, Nishinda/Buan - Shrub having bluish-purple flowers almost throughout the year.

লালচাঁদ The Crimson Rose
/*Tros nector Pachitopta nector*

ছোট লালচাঁদ The Common Rose
/*Tros aristolochiae*

সাদাকৈটা The Common Mormon
/*Papilio polytes*

নীলতিলক The Blue Mormon
/*Papilio polymnestor*

হলদে তিলক The Lime Butterfly
/*Papilio demoleus*

নীলকাট The Common Bluebottle
/*Zetides sarpedon*

বাবামি কাক The Common Indian Crow
/*Eupioea core*

নীলবাঘা The Blue Glassy Tiger
/*Danais limniace*

বেডোরা বাঘা The Plane Tiger
/*Danais chrysippus*

গুলবাঘা The Common Leopard
/*Atella phalantha*

বিষরূপ The Common Jezebel
/*Delias eucharis*

ডিমছাপ The Great Eggfiv
/*Hypolimnnas bolina*

শুখাপাতা The Common Evening Brown
/*Metanitis leda*

বেদিয়া The Common Emigrant
/*Catopsilie crocate*

হলুদ বেদিয়া
The Lemon Emigrant
/*catopsilia pomona*

কিন ডোরা The Common Sailor / *Neptis hylas*

পাহাড়িয়া ডোরা The Common Sergeant / *Pantoporia pertus*

মুসাফির The Common Wanderer / *Pareuonia valeria*

শ্বেত-কমলা The White Orange Tip / *Ixias marianne*

লীত কমলা The Yellow Orange Tip / *Ixias pyrene*

পোড়ামাটি The Tawny Coster / *Telchinia violae*

হলুদ মঞ্জিকা The Yellow Pansy/*Precis hieria*

নীল মঞ্জিকা The Blue Pansy / *Precis orithyia*

জালতি বা সাদা জালতি The Pion / *Anaphacis aurota*

ময়ূরকন্ঠী মঞ্জিকা The Peacock Pansy/*Piecisafmana*

নীল ঘাস পথিক The Pale Grass Blue/*Zizeeria maha*

ঘাসমণি The Grass Jewel / *Zizeeria putti (trochilus)*

ঢোলা The Common Gull / *Hupina or copora herissa*

হলদে ঘাস পথিক The Common Grass Yellow/*Terias hecube*

গোলাপ বালা The Painted Lady / *Vanessa cardul*

2.4 Indigenous Process for developing auspicious limestone dye

Abstract

Potential value of easily available natural limestone dyes used by the housewives, potters, masons for auspicious painting or marking, colouring walls or idols of god and goddess, etc. in Bengal have been discussed in this chapter for careful conservation of natural resources.

Keywords: Limestone dyes, Taldangra, Indigenous Process.

Introduction

Natural plant based dyes and clay based dyes have been in use since the time immemorial[1-2]. In West Rarrh the colour of the clay varies from white to brown to pink or black. The action of forces produced by nature, the duration of exposure and vegetation generally determine the types of soil in any region. Generally the top soil which are essential for plants may be classified as: Gravelly soil, sandy soil, clay soil, silty soil, loamy soil, calcareous soil, saline soil, laterite or red soil, black soil, humus soil etc. Due to intensive cultivation adopted, saving the soil for future use has generally been of decreasing concern over the last four to five decades. Besides many young people do not take farming these days due to lack of dignity of labour. So, some plants indicators like kendu (*Diospyros exsculpta*), *Spermacoce stricta* etc., which are vigorously grown on limestone soil and low cost soil testing kits etc. will be able to test sample in different lateritic belt of West Rarrh for proper using the soil resources wisely in view of eco-friendly development as well as easy steps for conserving the soil for future. After careful observation of soil profile it is found that a vast region of Taldangra Block under Bankura district of West Rarrh region contains limestone mines or pits. There are so many limestone refinery situated in Saltora, Ratanpur, Golkunda, Monipur, Kadagot villages of Bankura district. Natural soil based dyes have been in use since 500 years and the age of the limestone is approximately 70 million years. The limestone occurs at a depth from 1-10 m in the laterite belt. Nearly one lakh people are engaged in limestone dye industry. In unpartitioned Bengal the limestone dye used as a primer dye on the mud made idols of Devi Durga is claimed to be first propounded by the great poet Jagatram Roy inhabitant of village Bhului under P.S. – Mejia of Bankura district.

Materials and Methods

For preparing the dye Limestone is collected from pits. Usually 500 kg raw Limestone yields 75-80 kg dye. For boiling, Limestone is put into a covered earthen vat along with sufficient amount of rain water. In the next phase, in another vat limestone is stirred manually by legs with the help of water followed by precipitation. In the third phase of processing the upper supernatant solution along with silt particles is passed out in the third vat through a connecting link and kept in sunlight for 7 days. White coloured clay cakes gets deposited in the inner wall of the vat which is the instant for use as clay dye. The shelf-life of the dye is quite stable at the temperature ranging from -2 to 21°C. The cost of one ton dye is Rs. 1500/-.

Results and Discussion

These dyes are being transferred from limestone industries to Kumartuli-the famous place for idol making. The dye used as a primer diminishes the cost of valuable dyes. But at present due to increase in expenses for making this natural dye compared to synthetic dyes the dye-makers are facing in distress. The great advantage of this dye is that it has no adverse side effect. So, it is claimed for further research to meet the demand of numerous festivals of Bengals.

References

1. Hooper D., Some instances of vegetable pottery, J Asiatic Soc, 1906, **2** (3), 65-67.

2. Ghosh A., Plant and clay dyes used by weavers and potters in West Bengal, J Nat Prod Rad, 2004, **3** (2), 91.

2.5 Traditional Knowledge on auspicious plants for ancestral abode in ecological perspective

Abstract

Facets of indigenous knowledge on prosperous and evil plants for ancestral abode in ecological perspective are discussed. A revolutionary study and experience of the   eco-club for folklore studies is presented in documenting and promoting the eco-friendly homesteadgarden.

Keywords: Indigenous knowledge, West Rarrh,   Eco-friendly

Introduction

The West Rarrh mainly constituted by 7 districts such as Bankura, Birbhum, Burdwan, Purulia, Murshidabad, Purba and Paschim Medinipur. It is the extended part of the Chhotanagar plateau. It shows the gradual slope from north to south. Here the climate is basically hot and humid with a short period of winter. Soils are mainly lateritic, red and alluvial. Here tribals generally build small huts and their walls are constructed by plastering mud on branches of plants or plant parts. Both sides of walls are polished with fruit shells of *Bauhinia vahlii* and coloured with particular variety of clay after mixing with roots barks *decoction of Cleistanthus collinus* (parashi) and decoction of fruits of *Diospyros melanoxylon* (kend) and *Semecarpus anacardium* (bhelai) for protection of walls and thatched materials from white ants.

Traditional knowledge of the primitive men about their surroundings is still very effective to combat as well as to survive with the nature in the remote village (Banerjee 1974, Chaudhuri and Pal 1975). In Vedic era 12 Aranyas were identified (*viz* Dandyaka,Vindhya, Puskara, Nomisharanya, Kurujangala, Utpalabartakaranya, Jambukaranya, Arbudaranya, Himbadaranya, Dharmaranya, Vedaranya, Saindhabaranya) where they adapt themselves and modify the physical environment so as to reduce the limiting effects of physical conditions of existence. Tribals have number of evil spirits which must be propitiated by means of sacrifices. It is fear, not love or any reverential feeling, which prompts them to worship these spirits. They are very much afraid of them and they try with their idea of the Supreme Being (God), he is kind enough and understanding. Evil spirits (bhuts) and witches (dains) are responsible for diseases, troubles and ill fates and the best way to overcome them is to have recourse to the witch-finder. Their close association with various plants can be envisaged in their practices as describe here.

Materials and Method

Several field survey was done by making rappot with the knowledgeable men. Plants specimens were collected from the ancestral abode, authenticated and kept in the Herbarium of the Institute. On ecological perspective the auspicious plants and evil plants is given separately in Table 2.2.

Results and Discussion

The data has been accrued from the tribal people of the 7 districts which reflect their close affinity with the surrounding plants and their traditional way of life. However, isolation of effectiveness on either prosperous or evil plants, investigation is desired to validate the claims which may be further encouraged for the education for sustainable development. Though in the tribal areas the rules and regulation by which the tribes have been traditionally governed are now being gradually abolished and replaced by the young literate generations. While conducting the survey it is revealed that most of the people were dependent on plants by their local cultural faith and on traditional myth. Besides tribes have a myth that one large tree is ecologically equivalent to ten sons. Sons can betray with their parents, but this is not applicable for economically important plants. Here till now tree represents like 'The supreme natural symbol' of dynamic growth, seasonal death and regeneration. There are so many 'Sacred groves' in West Rarrh where as well as in the homestead garden trees worshipped as 'Totem'. Plants can absorb as well as resist the emission of harmful wave, radiation, dust etc. Though according to the ecologists big tree lying nearby homestead garden may induces the crack in the huts as well as enhances the emission of CO_2 which is deleterious to health. For ancestral abode ecologists classified the cultivated plants into 3 groups: (a) Vanaspati *i.e.,* non flowering plants like tree-ferns, satamuli etc. (b) Herbs like *Piper betle,* tulsi, money plant etc. (c) Shrubs like dalim, tagar etc.

Table – 2.2

(a) Prosperous Plants for ancestral abodes are:–

1. Supari (Betel-nut, *Areca catechu* L., Arecaceae)

2. Nagkeswar (Iron wood, *Mesua nagassarium* Burm.f., Hypericaceae)

3. Narikel (Coconut, *Cocos nucifera* L., Arecaceae)

4. Bel (Wood-apple, *Aegle marmelos* L., Rutaceae)

5. Dalim (Pomegranate, *Punica granatum* L., Punicaceae)

6. Ashoke (Asoka, *Saraca indica* auct., Caesalpiniaceae)

7. Tulshi (Sweet basil, *Ocimum basilicum* L., Lamiaceae)

8. Krishnatulshi (Sacred basil, *Ocimum sanctum* L., Lamiaceae)

9. Golap (Rose, *Rosa indica* L., Rosaceae)

10. Karabi (Oleander, *Nerium indicum* Mill., Apocynaceae)

11. Sajina (Drum stick, *Moringa oleifera* Lamk, Moringaceae)

12. Shet akanda (*Calotropis gigantea* L., Asclepiadaceae)

13. Jamrul (Water rose-apple/star apple, *Syzygium samarangensa* Blume, Myrtaceae)

14. Sthalpadma (Changeable rose, *Hibiscus mutabilis* L., Malvaceae)

15. Tagar (Wax flower, *Tabernaemontana divaricata* L., Apocynaceae)

16. Am (Mango, *Mangifera indica* L., Anacardiaceae)

17. Kala (Banana, *Musa paradisiaca* L., Musaceae)

18. Golap jam (Rose apple, *Eugenia jambos* L., Myrtaceae)

19. Chameli (Spanish jasmine, *Jasminum grandiflorum* L., Oleaceae)

(b) Evil plants for ancestral abodes are :

1. Neem (Margosa, *Azadirachta indica* A. Juss., Meliaceae)

2. Palash (Flame of the forest, *Butea monosperma* Lamk., Fabaceae)

3. Shimul (Silk cotton, *Bombax ceiba* L., Bombacaceae)

4. Khejur (Date palm, *Phoenix sylvestris* L., Arecaceae)

5. Jagnya dumur (Fig, *Ficus racemosa* L., Moraceae)

6. Raktakanchan (Camel's fool, *Bauhinia purpurea* L., Caesalpiniaceae)

7. Ketaki/Keya (Screw-pine, *Pandanus fascicularis* Lamk., Pandanaceae)

8. Tentul (Tamarind, *Tamarindus indica* L., Caesalpiniaceae)

9. Kalojam (Black plum, *Syzygium cumini* L., Myrtaceae)

10. Bot (Banyan tree, *Ficus benghalensis* L., Moraceae)

11. Ashwatha (Peeple, *Ficus religiosa* L., Moraceae)

12. Lebu (Orange, *Citrus medica* L., Rutaceae)

13. Arjun (White murdah, *Terminalia arjuna* Roxb., Combretaceae)

14. Kul (Plum, *Ziziphus mauritiana* Lamk., Rhamnaceae)

15. Pakur (*Ficus infectoria* Roxb., Moraceae)

16. Kadam (*Anthocephalus cadamba* Roxb., Rubiaceae)

17. Malati (*Aganosma caryophyllata* Roxb., Apocynaceae)

18. Mallika (Arabian Jasmine/Tulip, *Jasminum sambac* L., Oleaceae)

19. Dhutura (Thorn apple, *Datura metel*, L., Solanaceae)

20. Bahera (Beleric, *Terminalia bellirica* Gaertn., Combretaceae)

21. Haritaki (Black myrobalan, *Terminalia chebula* Gaertn., Combretaceae)

22. Bans (Bamboo, *Bambusa arundinacea* Retz., Poaceae)

References

1. Banerjee DK, Magico-religious beliefs about plants among some Adivasis of India, J. Mythic Soc. 1947, **65** (3), 5-8.

2. Chaudhuri Rai HN, and Pal DC, Notes on magico-religious belief about plants among Lodha of Midnapore, West Bengal, Vanyajati, 1975, **23** (2-3), 20-22.

2.6 Mosquitovorous Plants of West Rarrh

Abstract

The efficacy of some plants against mosquito is surprising as they served as a natural potential source which have no adverse side effect.

Keywords : West Rarrh, Mosquitovorous, Junglemahal.

Introduction

West Rarrh is endowed with a rich ethnobotanical potential. For which ancient tribes (*viz.* Santal, Lodha, Kheria, Bhumij etc.) can survive since the time immemorial without allopathic medicine. So, it sometimes claimed that folk medicine have been handed down from the God. Most of the tribes in the Junglemahal area of West Rarrh have no mosquito-curtain and they live with their body's own defence mechanism. Though the "Asian tiger mosquito" *Aedes albopictus* is an important vector of dengue is flying here and there. But some plants can combating with the mosquito through natural process, thereby they maintains the process of biological control. So they can compensate the role of synthetic mosquitocides without having any resistance developed by the mosquitoes.[1-2]

Materials and Methods

The present work was carried out in many wetlands in the year 2001 and the botanical identity of plants was authenticated by an eminent taxonomist (by Prof. M. N. Sanyal) and the voucher specimens were deposited in the herbaria of Swamiji Eco-club for future use.

Results and Discussion

All the plants exhibited significant mosquitovorous property. Further research in this pursuit promises to open new findings in the use of these plants. Therefore, it seems more feasible to use these plants for commercial purpose in the wetland management system where they may be attained the status of beneficial weeds and thereby maintained the ecological balance.

References

1. Brown AWA, Insecticide resistance in mosquitoes: a pragmatic review, J Am Mosq control Assoc, 1986, **2** (2), 123 – 139.
2. WHO, Vector resistance to pesticides, Fifteenth report of the WHO expert committee on vector biology and control, WHO Tech Rep Ser, 1992, **818**, 1 – 62.

Table 2.3 : List of Mosquitovorous Plants

Sl. Botanical Name/Collection No.	Family	Local Name	Occurrence/Habitat
1. *Chara zeylanica* Willd./Ghosh 600	Characeae	Stone-wort	Found in ponds.
2. *Drosera burmannii* Vahi/Ghosh 105	Droseraceae	Surjasisir	Common in streamlets, rice fields.
3. *Geranium zonale* Linn./Ghosh 701	Geraniaceae	Dora	Found in garden.
4. *Nitella acuminata* Braun/Ghosh 606	Characeae		Found in ditches.
5. *Utricularia aurea* Lour./Ghosh 309	Lentibulariaceae	Jhanji (Bladder wort)	Abundant in tanks and pools.
6. *Utricularia bifida* Linn./Ghosh 310	Do	Do	Occurs in marshy places.
7. *Utricularia caerulea* Linn./Ghosh 311	Do	Do	A weeds of rice fields and swampy places.
8. *Utricularia exoleta* R. Br. /Ghosh 312	Do	Do	In swampy places.
9. *Utricularia stellaris* Linn./Ghosh 313	Do	Do	Traps numerous, occasional in pools and rice fields.

2.7 Cult of the Flower Deities of Bengal

One of the most mysterious creature of nature is the flower. Most of the man love flowers for the beauty as well as for their scented smell. So, the people have a common myth that different Dev-Devi have different fond for the flowers. Therefore for the rituals of God and Goddess the foremost requirement is the flower. In the scarcity of flower leaf – fruit – kush (*Desmostachya bipinnata* L., Poaceae) may be used. Generally abscised flowers are prohibited for these purpose except sheuli and bakul. Similarly state flowers and buds of lotus are not acceptable.

Ritualistic competeney : Kherias, Lodhas and Mahali have some evil spirits commonly called Bhuts. It is fear, not love or any reverential feeling, which prompts them to worship these spirits. Evil spirits (bhuts) and witches (dauns) are responsible for troubles and diseases and the best way to get over them is to have recourse by means of sacrifices. In the spring festival, the sacrifice takes place at Sacred Grove. In this festival flowers of Mahua (*Madhuca indica*), Dhawai (*Lagerostomea parviflora*) and

Table 2.4 : Duration of ritualistic competency of different flowers

Local Name	Botanical Name with family	Duration (in days)
Padma	*Nelumbo nucifera* Gaertn. Nymphaeaceae	10
Chameli	*Jasminum grandiflorum* L. Oleaceae	2
Champa	*Michelia champaca* L. Magnoliaceae	5
Palash	*Butea monosperma* (Lamk.) Taub, Fabaceae	7
Karabi	*Nerium indicum* Mill. Apocynaceae	8
Ketaki (Keya)	*Pandanus fascicularis* Lamk. Pandanaceae	5
Dalim	*Punica granatum* L. Punicaceae	5
Sal	*Shorea robusta* Gaertn. Dipterocarpaceae	5
Jaba	*Hibiscus rosa – sinensis* L. Malvaceae	2
Gandharaj	*Gardenia jasminoides* Ellis Rubiaceae	4
Tagar	*Tabernaemontana divaricata* L. Apocynaceae	4
Kunda	*Jasminum multiflora* Andr. J. pubescens* Willd. Oleaceae J. arborecens* Roxb.	2

sal (*Shorea robusta*) gathered for worshipping in the sacred grove. Besides mango flowers (*Mangifera indica*) are offered.

Name of God	Name of flower which offer	
& Goddess	Local Name	Botanical name with family
Bishnu	Brahma Kamal	
Siva	Akand Dhutra Sivalinga	*Colotropis gigantea* (L.) R. Br., Asclepiadaceae *Datura metel* L., Solanaceae *Couroupita guianensis* Aubl, Lecythidaceae
Saraswati	Shet padma Palash Atang Am	*Nelumbo nucifera* Gaertn, Nymphaeaceae *Butea monosperma* (Lamk) Taub, Fabaceae *Combretum decandrum* Roxb, Combretaceae *Mangifera indica* L, Anacardiaceae
Narayan	Karabi	*Nerium indicum* Mill, Apocynaceae
Chandi	Karabi Rajanigandha	Do *Polianthes tuberosa* L, Amaryllidaceae
Manasa	Padma	*Nelumbo nucifera* Gaertn.
Kali	Jaba	*Hibiscus rosa – sinensis* L. Malvaceae

3. For Human Purposes

3.1 Folk medicine of West Rarrh from Aquatic Plants

Abstract

The present scientific communication focuses upon the first hand information of the tribal knowledge related to the healing property of 38 aquatic plant species belonging to 22 families with their local names and ethnomedicinal uses to cure different aliments associated with tribal people of West Rarrh.

Keywords: Aquatic plants, West Rarrh, Synthetic drug.

Introduction

Plant therapy is quite prevalent in remote tribal villages where hospital facilities are rarely available. So, the traditional ethnic knowledge of aquatic plants proves useful among the aboriginals. The West Rarrh is situated at a longitude of 85°70' to 87°80' E, latitude of 21°75' to 24°33' N and altitude ranges from 50 metres to 750 metres above mean sea level. West Rarrh is the home of many tribals (*viz.* Gonds, Kols, Mahali, Sabar, Lodha, Mech, Munda, Santal, Oraon, Bhumij, Bedia, Birhore etc.) and rural people mostly living in remote areas chiefly rely upon herbal medicine rather than going to 'Primary Health Care Centres' because of the awareness of the side effects and toxicity associated with the synthetic drugs and their unavailability. Throughout the approaches West Rarrh so many lentic and lotic water bodies are present harbouring vigorous growth of aquatic plants which are the chief sources for food, fodder, fuel, medicine etc. Medicinal efficacy of aquatic plants are recorded by different workers. Review of the literature revealed that West Rarrh is unexplored ethnobotanically. The present paper is the first attempt to unveil the medicinal value of the aquatic herbs. Besides there is urgent need to document and restore the aquatic bodies for water conservation.

Materials and Methods

A regular planned field expedition (during 2002-2005) were made to local ethnic healers at different seasons and the ethnobotanical information was collected through interrogation. Before the information was recorded it was repeatedly verified. Plant species used by the tribals are listed in the Table 1 with their botanical names, specimen no. and local names, family, parts used and mode of recipe and its uses. Plant specimens were collected from the study area, identified by Prof. M.N. Sanyal (Head of the Dept. of Botany, Bishnupur R. N. College) and kept in the Herbarium of the Eco-club for future use. The uses marked with the asterisk (*) are the new information's (as hither to not known or uncommon in the study area).

Results and Discussion

During the surveys it has been observed that inhabitants of West Rarrh still depend upon the aquatic herbs for curing different ailments like cold, cough, fever, stomatitis, diarrhoea, piles, dentitis etc. Of the total 38 plant species encountered maximum number of plant species are being used in fever. Further, it has been observed that leaves are frequently used against various ailments. Most of the recipe are in the form of fresh extract. The survey revealed that aquatic herbs are highly diverse and possessed enormous ethnomedicinal resource potential. Hopefully, some of the new information's may lead to development of commercial products after phytochemical and pharmacological investigations.

References

1. Biswas K and Calder CC, Hand book of Common Water and Marsh Plants of India and Burma, Govt. of India Press, Calcutta, 1954.

Table 3.1 : List of useful aquatic plants

Botanical name/Collection No.	Family	Local name	Parts used	Mode of recipe and its use's
1. *Acorus calamus* L./Ghosh 12	Araceae	Pani boch	Whole plant, Rhizome	Decoction　　　　is used in pox, cold, cough, epilepsy, fever, stomatitis.
2. *Atisma plantago* L./Ghosh 18	Alismataceae	Water plantain	Leaf	Powdered leaves are used externally in itching.
*3. *Azolla pinnata* R br./Ghosh 22	Azollaceae	Khudi pana	Leaf	Dry leaves (25 g) with turmeric powder (2g) are used in food twice daily for egg production in domestic birds.
4. *Ceratopteris thalictroides* L./Ghosh 83	Parkeriaceae	Dhenki shak	Leaf	Juice extract　　　is used in fever once daily.
5. *Coix aquatica* L./Ghosh 99	Poaceae	Job's tear	Leaf	Do
*6. *Commelina benghalensis* L./Ghosh 103	Commelinaceae	Kanshira	Leaf	Fresh leaf extract is used as eye drop thrice daily against cataract.
7. *Eichhornia crassipes* Marr./Ghosh 133	Pontederiaceae	Kachuri pana	Flower, Root	Paste of the flower is used to treat gout. Root paste is used in toothache.

Botanical name/Collection No.	Family	Local name	Parts used	Mode of recipe and its use's
*8. Enhydra fluctuans Lour./Ghosh 135	Asteraceae	Hingcha	Leaf	Juice is used once daily for high blood pressure, as blood purifier and in diabetes.
9. Euryale ferox Salish/Ghosh 145	Nymphaeaceae	Makhna	Seed	Used 25g seeds regularly in general weakness.
*10. Hydrilla verticillata L./Ghosh 171	Hydrocharitaceae	Jhanji	Whole plant	Intake of plant extract improve blood circulation and check the diabetes.
11. Hygrorhiza aristata Retz./Ghosh 172	Poaceae	Jangli dal	Root	Root juice (5 ml) is used thrice daily in diarrhoea.
12. Imperata cylindrica Beauv./Ghosh 173	Poaceae	Ulu	Root	Root juice is used externally in ring worm.
13. Ipomoea aquatica Forsk./Ghosh 175	Convolvulaceae	Kalmi shak	Leaf	Fresh juice (1 cup) is used thrice daily as lactogenic.
*14. Lemna trisulca L./Ghosh 188	Lemnaceae	Pana	Whole plant	Plant paste mixed with Aloe vera extract alongwith coconut oil is smeared over head in case of baldness.

Botanical name/Collection No.	Family	Local name	Parts used	Mode of recipe and its use's
15. *Limnophila conferta* Benth./Ghosh 190	Scrophulariaceae	Hemcha shak	Leaf	Juice (½ cups) is consumed orally in fever twice daily.
16. *Marsilea quadrifolia* L./Ghosh 198	Marsileaceae	Susni shak	Leaf	The decoction of leaves (½ cups) along with ginger (*Zingiber officinale*) is used to cure cough and bronchitis.
17. *Monochoria vaginalis* Presl./Ghosh 201	Pontederiaceae	Kachuripana	Leaf	Leaves are used (1 cup) as lactogenic.
*18. *Nelumbium nucifera* Gaertn./ Ghosh 214	Nymphaeaceae	Padma	Seed, Leaf, Flower	(a) Drunk the paste of 4 seeds with ½ glass water for 21 days in habitual abortion. (b) Decoction of leaves (5 ml) is used twice daily in jaundice and as anti HIV. (c) Flowers (5 petals) are used as cardio tonic and styptic during gestation.

	Botanical name/Collection No.	Family	Local name	Parts used	Mode of recipe and its use's
19.	*Nymphaea lotus* f./Ghosh 218	Nymphaeaceae	Shalook	Root, Seed	a) Paste of 5 roots with cup milk is taken orally twice daily for 14 days in piles. (b) Fried seeds (10 g) are used in constipation.
*20.	*Nymphaea nouchali* Burm. f./ Ghosh 220	Do	Red shalook	Flower, Rhizome	a) Juice of flower is used for 3 months to achieve impregnation. (b) Extraction of rhizome along with 200g curd is used as a cold drink in diarrhoea.
21.	*Nymphaea rubra* Roxb./Ghosh 221	Do	Lal padma	Leaf	Juice is taken against piles.
22.	*Nymphoides cristatum* Roxb./ Ghosh 222	Gentianaceae	Jal phool	Leaf	Decoction of leaves is used twice daily in jaundice.
23.	*Nymphoides indicum* L./Ghosh 223	Do	Do	Rhizome	Decoction of rhizome is used in fever.

Botanical name/Collection No.	Family	Local name	Parts used	Mode of recipe and its use's
*24. Oryza rufipogon Griff./Ghosh 226	Poaceae	Basudhan	Whole plant	Plant extract is used orally thrice daily in conjunctivitis.
*25. Oryza sativa L./Ghosh 227	Poaceae	Kaya dhan	Endosperm	1 cup gruels (endosperm boiled in water) of the unpolished rice is used for lowering the blood cholesterol.
26. Pistia stratiotes L./Ghosh 230	Araceae	Toka pana	Leaf	Extract of leaves is taken against ring worm.
27. Polycarpon loeflingiae Benth./Ghosh 240	Caryophyllaceae	Gima shak	Leaf	Juice is used in fever twice daily.
28. Polygonum hydropiper L./Ghosh 245	Polygonaceae	Pakurmul	Root, Leaf	(a) Smear the root paste in dentitis. (b) Leaf extract along with honey is consumed orally to kill hook worms.
29. Polygonum plebejum R. Br./Ghosh 246	Do	Do	Leaf	Juice is used in fever.
*30. Potamogeton crispus L./Ghosh 250	Potamoge-tonaceae	Bans khagra	Leaf	Extract of leaves are taken in diabetes.

Botanical name/Collection No.	Family	Local name	Parts used	Mode of recipe and its use's
31. *Sagitaria sagitifolia* L./Ghosh 265	Alismataceae	Chotokut	Leaf	Leaf juice is applied externally in rashes and allergies.
32. *Salvinia cucullata*/Ghosh 267	Salviniaceae	Indurkani	Leaf	a) Leaf juice (5 ml) is taken in dysentery. (b) Dry leaves (250 g) is lactogenic for domestic animals.
*33. *Scirpus grossus* L./Ghosh 268	Cyperaceae	Patpati	Tuber	Paste of tubers (50 g) are taken in diabetes.
34. *Spirulina platensis* Nordst./Ghosh 271	Oscillatoriaceae	Neel-sheula	Whole plant	Plant extract used as vitamin tonic and sex promoter.
*35. *Trapa bispinosa* Roxb./Ghosh 302	Trapaceae	Paniphal	Fruit	Fresh fruits are used as aphrodisiac.
*36. *Typha elephantina* Roxb./Ghosh 310	Typhaceae	Hogla	Root	Boiled root (20 g) is used orally in spermatorrhoea.
37. *Utricularia exoleta* R. Br./Ghosh 312	Lentibulariaceae	Bladderwort. (Jhanji)	Plant extract	Used as tonic in fever.
*38. *Vallisnaria spiralis* L./Ghosh 315	Hydrocharitaceae	Patajhanji	Leaf	Paste of the leaves (25 g) with 15 g *Foeniculum vulgare* is given to the child passing green stools.

2. Subramanyan K, Aquatic angiosperms, CSIR, New Delhi, 1962.

3. Majid FZ, Aquatic Weeds, Utility and Development, Agro Botanical Publishers, (India), 1986.

4. Gupta A, Sahoo TR and Tiwari E, Ethnomedicinal importance of some common aquatic and marshy plants of Sagar district, J. Bot, Soc, Uni, Sagar, 2005, 4, 65 – 73.

5. Syeeda M, Anand VK and Shah NH, Ethno-medico-botany of some important aquatic plants of Jammu Province (J & K) India, J. Phytol Res, 2008, 21, 269 – 272.

6. Baro S P, Balarabe M L and Ruga BT, Ethnobotanical, Economic and Nutritive Potentials of Aquatic Vascular Flora in the Nigerian Savanna, In : Recent Progress in Medicinal Plants (Vol. I Ethnomedicine and Pharmacognosy) by V K Singh, J.N. Govil and G. Singh, (Eds) SCI Tech Publishing LLC, USA, 2002, pp 177–182.

7. Rao R R and Majna P K, Methods of Research in ethnobotany, In : A Manual of ethnobotany by S.K. Jain, Scientific Publisher, Jodhpur (India), 1987, pp. 83–91.

3.2 Herbal remedies for alopecia used by the tribals of West Bengal

Abstract

Use of alopecia recovering plants is a common practice among the tribals. In this chapter a list of 31 angiospermic plants species belonging to 22 families used against alopecia. The study was performed within the tribals of Kheria, Lodha, Santal, Munda, Kurmi, Bhumij Oraon, Birhore and Jatorh. Among these, Keria and Santal are more knowledgeable than others about the plants usage. The mode of preparation, administration and the dosage are provided in the table.3.2

Keywords : Alopecia, Ethnomedicine, Herbal remedies, Sacred groves, WBC.

Introduction

West Bengal is endowed with natural abundance of diverse flora including an enormously large number of ethnomedicinal plants. During present survey an attempt has been made to collect correct information from different tribes. The available literature dealing with ethnomedicines have been consulted with the field study.

Patches on the scalp with absolutely no hair is called Alopecia (*A. areata*). In *Alopecia areata*, the affected hair follicles are affected by a person's own immune system (WBC), resulting in the arrest of the hair growth stage. Though complete body hair loss is taken place by *Alopecia universalis.*

In *A. universalis*, however loss of eyelashes, hair in the nose as well as in ears make the man more unprotectable to dust, germs etc.

Study area

West Bengal is well diversified along with unique vegetation due to varied geographic and climatic condition. It has 11,548 Km^2 forest area (*i.e.* 13%) of the total land. The major vegetation at the state are: dry tropical forest, sub tropical forest, moist tropical forest, temperate forest and grass land.

The West Bengal lies in between 21°45' to 27°10' N latitude and 85°55' to 89°56' E longitude, comprising of an area of about 88,752 km^2.

Still, tribals are dispersed in small pockets where adequate modern medical facilities are not available. They solely depend on ethnomedicine which are generally available on sacred groves.

Materials and Methods

During the survey all the above mentioned tribes are interrogated and documentation of their indigenous knowledge and plants used are presented. Two major groups of people, *viz.*, Kurmi and Jatorh are not tribe but have profound knowledge as they residing in the forest area. Voucher specimens are preserved for future. The uses marked with the asterisk (*) are the new information. Plant species used by the tribals are listed in the table with their botanical and vernacular names, family, parts used and mode of uses.

Table 3.2 : Antialopecial plants used by the tribals

Botanical Name	Family	Local name	Parts used	Mathod of Use
1. *Abrus precatorius* Linn.	Fabaceae	Kunch	Ripe seed	**Seed paste rubbed on the scalp twice daily.**
2. *Acacia concinna* DC.	Mimosaceae	Shikakai	Pods extract	**Pod extract is applied on the scalp once daily.**
*3. *Allium cepa* Linn.	Liliaceae	Pianj	Bulb	**Juice of squeezed bulb is applied on scalp.**
4. *Arnica montana* Linn.	Asteraceae	Arnica	Flower	**Flower extract is used in alopecia.**
5. *Bacopa monnieri* Linn.	Scrophulariaceae	Brahmi	Leaf	**Leaves are boiled in seasame oil and then massaged on scalp.**
*6. *Bambusa arundinacea* Retz.	Poaceae	Bans	Root	**Smear the ash of root with sesame oil.**
*7. *Caesalpinia crista* Linn.	Caesalpiniaceae	Nata Karanj	Seed oil	**Smear the oil twice daily.**
8. *Camellia sinensis* Linn.	Ternstroemiaceae	Cha	Tender leaf	**Boiled leaf extract cheek the hair loss.**
9. *Centella asiaica* Linn.	Apiaceae	Thankuni	Leaf	Smear the leaf juice on scalp.
*10. *Citrullus colocynthus Schrad.*	Cucurbitaceae	Makal(Tito sasha)	Root	Fresh root paste used thrice daily.
11. *Coriamdrum sativum* Linn.	Apiaceae	Dhoney	Seed	Fresh seed paste with sesame oil used daily.

Botanical Name	Family	Local name	Parts used	Mathod of Use
*12. Cynodon dactylon Linn.	Poaceae	Duba	Plant	Used the plant paste on this scalp with sesame oil.
13. Datura metel Linn.	Solanaceae	Dhatura	Root	Rub the ash of root with sesame oil.
14. Eclipta alba Linn.	Asteraceae	Keshute	Plant	Used the plant paste on scalp with sesame oil.
*15. Euphorbia nerifolia acut.	Euphorbiaceae	Sij – monsa	Latex	Smear the fresh latex along with sesmae oil.
16. Hibiscus rosa – sinensis Linn.	Malvaceae	Jaba	Flower	20-25 fresh flowers are crushed and boiled in 200g sesame oil for daily usage on the scalp.
17. Lawsonia inermis Linn.	Lythraceae	Mehendi	Leaf	Paste of leaves used daily on the scalp.
*18. Momordica cochinchinensis Lour.	Cucurbitaceae	Kankrol	Root	Used the fresh root paste on the scalp.
19. Piper nigrum Linn.	Piperaceae	Golmarich	Fruit	Fruit containing seed paste along with a pinch of rock salt smear on the scalp.
20. Portulaca oleracea Linn.	Portulacaceae	Nunia	Whole plant	Extract claimed as recovery of hairs.
21. Pyrus communis Linn.	Rosaceae	Nashpati	Fruit	Fruit juice used against alopecia.

Botanical Name	Family	Local name	Parts used	Method of Use
22. *Ricinus communis* Linn.	Euphorbiaceae	Rerhi	Seed oil	Massaging the oil to prevent hair loss.
23. *Sesamum indicum* Linn.	Pedaliaceae	Til	Seed	Oil is used in alopecia.
24. *Solanum indicum* Linn.	Solanaceae	Brihati	Fruit	Smear the juice with honey.
25. *Swertia chirata* Ham.	Gentianaceae	Chireta	Plant	Use the leachate of plant on the scalp.
26. *Terminalia bellirica* Roxb.	Combretaceae	Bahera	Fruit	Smear the fresh fruit pulp on scalp.
27. *Tribulus terrestris* Linn.	Zygophyllaceae	Gokhur	Seed	Seed paste along with sesame flower paste mixed with ghee and honey smear on the scalp.
*28. *Vetiveria zizanioides* Linn.	Poaceae	Bena	Root	Fresh root extract used on scalp
29. *Vitex negundo* Linn.	Verbenaceae	Buan/Nishinda	Leaf	100 g leaves are boiled in seasame oil and use for massaging on scalp to prevent hair loss.
30. *Wedelia calendulacea* Less.	Asteraceae	Bhringaraj	Extract of leaves	Check the hair loss.
31. *Zea mays* Linn.	Poaceae	Bhutta	Stigma	Stigma extract used as lotions for alopecia.

Results and Discussion

Among the tribal communities, the Kherias and the Santals are the most well acquainted with the anti-alopecial drugs. Besides it is significant that some of the drugs are very effective in close contact of the sesame oil. So for achieving long beautiful, silky, bouncing and dark black hairs the above mentioned tips can be accepted easily by anybody as Alopecia is a highly unpredictable, auto immune skin disease resulting the hair fall. It is significant that alopecia drugs are rarely prepared from stigma and latex. Hopefully, some of the new informations may lead to development of commercial products as majority of the claims are used based on single plant parts.

References

1. Jain S K (ed), Glimpses of Indian Ethnobotany, Oxford IBH, Calcutta, 1981.

2. Jain S K, Manual of Ethnobotany, Scientific Publishers, Jodhpur, 1987, 26.

3. Ghosh A, Maity S and Maity M, Ethnomedicine in Bankura Dist. West Bengal, *J Econ Taxon Bot*, Addl Series, 1996, 12, 329-331.

4. Pal DC and Jain SK, Tribal Medicine, Naya Prakash, Calcutta, 1999.

5. Ghosh A, Ethnomedicine for Human and Veterinary Development, Daya Publishing House, Delhi, 2009.

6. Ghosh A, Ethnobiology: Therapeutics and Natural Resources, Daya Publishing House, Delhi, 2009.

4. For Veterinary Purposes

4.1 Ethno-veterinary medicine as practiced by the tribals of West Rarrh

Abstract

Present communication brings to knowledge the indigenous methods of treating sheep ailments using herbal drugs reported from rural folks in West Rarrh region of West Bengal. A total of 37 plant species belonging to 28 families of angiosperms are employed by the inhabitants in the form of decoction, infusion, oil, paste, latex etc. either as a solitary drug or in combination to treat various aliments. The doses, duration, method of administration etc. are however require further testing.

Keywords: Ethno-veterinary medicine. West Rarrh, Zoopharmacognosy, Ecobolic.

Introduction

Plants have been used as a source of veterinary medicine during the period of Charaka and Jibaka (Rajbaidya of Bimbisar and Lord Budhya; also a disciple of Atraya muni in Takhyasila Viswavidyalaya). According to an estimate of West Rarrh approximately 85 per cent of the people rely chiefly on indigenous knowledge for primary health care of the sheep. The literature survey shows that the region was almost unexplored. The present work focused on some more plant species from this region which are very effective and cheap. Generally, the tribals have the tendency to live together and always keep themselves busy in constructive activities and are faithful.

Study area

The tribals of West Rarrh are the Gonds, Kols, Mahali, Sabar, Lodha, Munda, Santal, Oraons, Birhore, Bhumij, Mech, Bedia etc. Within this region the districts are Bankura, Birbhum, Burdwan, Midnapore, Murshidabad, and Purulia. It is the extended part of Chhotanagpur plateau. Here the forest lie scattered in small patches between latitude 21°75' and 24°33' N and longitude from 85°70' to 87°80' E. Here the hills are relic type. In West Rarrh 3 types of soil namely alluvial, red and laterite soil are existing. There is a dominance of hot and humid climate along with a short duration of winter. Besides temperature reaches its maximum (39.45°C in average) in the month of May and falls in the month of January (12.57°C).

Materials and Methods

Ethnobotanical fields survey was carried out during the years 2000 – 2002 among the tribal communities. Information collected from them at first and then applied to the sheep to test their effectiveness. Plant specimens were collected from the study area, authenticated and kept in the herbarium. The ethno-veterinary medicinal information is enumerated in Table 4.1

Results and Discussion

Ethnic people are knowledgeable and their world view about the sustainable life is now extensively studied. Perhaps the first that recognized the medicinal values of plants may be the birds and animals. Dogs, monkeys, cats, tigers, sheep and other animals prefer some plants and animals during some special bodily conditions which is termed as zoopharmacognosy. As for example in Kappa of Ethiopia at first the sheep discovered the seeds of coffee as a stimulant. All claims made by the ethnic people should be tested for their validity.

Table 4.1 : Medicinal plants used for sheep in West Rarrh region

S.No.	Botanical name and Family/Collection No.	Local Name	Ailments	Parts Used and preparation	Mode of administration
1.	Abelmoschus esculentus L. Malvaceae/Ghosh 1	Dhenrash	Lactogenic	Root	Hang a root in tail after delivery to get as usual milk.
2.	Acacia concinna DC Mimosaceae/Ghosh 8	Bon-ritha/ Sikakai	Udder lesions	Pod	Washes the udder and smear the paste with turmeric powder and pinch of common salts for 10 days.
3.	Acacia nilotica L. Mimosaceae/Ghosh 5	Babla	Foot and Mouth disease/ Khur poka	Stem bark	Smear the paste of bark in the affected part of leg twice daily.
4.	Achyranthes aspera L. Amaranthaceae/Ghosh 11	Apang	i) Acne in tongue	Root paste	Rubbed the paste in tongue along with turmeric twice a day.
			ii) Parturition	Root	Root is hung in tail for placental retention during parturition.
			iii) Snake bite	Root	Paste (10g) is used both orally and externally.
5.	Allium sativum L. Liliaceae/Ghosh 22	Rasun	i) Fever	Bulb juice 2g, 5g of turmeric powder & 5g leaf leachate of tobacco	Drunk ½ cup infusion twice daily alongwith mustard oil.

S.No.	Botanical name and Family/Collection No.	Local Name	Ailments	Parts Used and preparation	Mode of administration
			ii) Castrated wound	Bulb boiled in mustard oil.	Smear the oil in wound.
			iii) Dog-bite	Bulb	Smear the paste on the wound.
6.	*Aloe vera* (L.) Burm.f. Liliaceae/Ghosh 24.	Ghritakumari	Loosen teeth	Leaf	Apply the paste.
7.	*Anona squamosa* L. Anonaceae/Ghosh 34	Ata	Loss of hair	Leaf juice	Regular external application is advised.
8.	*Asparagus racemosus* Willd. Liliaceae/ Ghosh 40.	Satamuli	Fertilization	Root	Intake of 250 g root paste of satamuli along with 250 g bark paste of banyan tree (*Ficus benghalensis* L., Moraceae) for 4 days from the date of mating with sheep's fodder.
9.	*Avicennia officinalis* L. Avicenniaceae/Ghosh 43.	Kalo-bean	Lactogenic	Leaf	Fresh leaves 1-2 kg / day.
10.	*Azadirachta indica* A. Juss. Meliaceae/ Ghosh 201	Nim	Loosen teeth	Twig	Apply the paste with $KMnO_4$.

S.No.	Botanical name and Family/Collection No.	Local Name	Ailments	Parts Used and preparation	Mode of administration
11.	*Bambusa arundinacea* (Retz.) Will Poaceae/ Ghosh 45	Bans	Retention of placenta	Leaf caruneles.	Due to lack of myometrial contractions fetal membrane are not expelled within 6 hrs. As a consequence the membrane remain attached to the uterine As an Ecobolic substance (fodder) bamboo leaves promotes the expulsion of the contents of the uterus.
12.	*Cissus quadrangularis* L. Vitaceae/Ghosh 91	Harjora	Bone fracture	Stem	Luke warm paste of harjora with sij-monsa (*Euphorbia nerifolia* acut.) externally applied as balm.
13.	*Curcuma longa* L. Zingiberaceae Ghose 116	Halud	Exudation of blood from the nipple	Rhizome	A mixture of 50 g turmeric powder and 5 ml of dettol is fed to the sheep.
14.	*Cuscuta reflexa* Roxb. Cuscutaceae/Ghosh 117	Swarnalata	Dermatitis	Stem	Smear the aqueous extract to affected part twice daily
15.	*Datura metel*/L. Solanaceae/Ghosh 121	Dhatura	Swollen throat	Dry flower	One sun dried flower administered twice daily.

S.No.	Botanical name and Family/Collection No.	Local Name	Ailments	Parts Used and preparation	Mode of administration
16.	*Diospyros melanoxylon* Roxb. Ebenaceae/ Ghosh 129	Kendu	Diarrhoea	Fresh fruit juice	One cup juice along with cold water recommended thrice daily in 2:3.
17.	*Mallotus philippensis* Lam. Euphorbiaceae/ Ghosh 350	Kapila/Guri	Worm infestation	Young plant juice	Juice is given to the sheep along with fodder till get healed.
18.	*Mimusops elengi* L. Sapotaceae/Ghosh 204	Bakul	Repeat breeder (fail to conceive even after 2 successive inseminations)	Stem bark	Bark and fennel seeds (*Foeniculum vulgare* Gaertn.) leachate water (250 ml fed to the sheep twice daily for 7 days.
19.	*Musa paradisiaca* L. Musaceae/Ghosh 213	Kala	Repeated Miscarriage	Pseudo-stem	½ cup juice thrice daily along with the fodder.
20.	*Nerium oleander* L. Apocynaceae/Ghosh 215	Karabi	Yoke gall	Leaf	Smear the decoction twice a day for 3 days.
21.	*Piper nigrum* L. Piperaceae/Ghosh 240	Gol-morich	Flatulence	Seeds	15 seeds are crushed with a pinch of rock salt and mixed in 100 g molasses and fed to sheep thrice a day.
22.	*Plantago ovata* Forsk. Plantaginaceae/Ghosh 241	Isabgol	Recurrent breeding	Seeds	Fed to the sheep 50g seeds for 3 days along with castor oil in 1:3.

S.No.	Botanical name and Family/Collection No.	Local Name	Ailments	Parts Used and preparation\	Mode of administration
23.	*Ricinus communis* L. Euphorbiaceae/Ghosh 260	Rerhi	Oozes out of milk from the nipple.	Leaf	Smear the paste thrice daily
24.	*Saccharum officinarum* L. Poaceae/Ghosh 263	Akh	Diarrhoea	Bagasse	Intake of paste of bagasse(25g) along with barks of jam (*Syzygium cumini* L. Myrtaceae)
25.	*Solanum tuberosum* L. Solanaceae/Ghosh 279	Alu	Corneal opacity	Tuber	When sheep suffer from corneal opacity eyes become cloudy-white, watery then administers 2 drops of filtered juice thrice daily.
26.	*Solanum xanthocarpum* Schrad & Wendl. Solanaceae/Ghosh 280	Kantikari	Cataract	Whole plant juice	Puts a few drops of the juice in the eyes thrice daily.
27.	*Strychnos nux vomica* L. Loganiaceae/Ghosh 285	Kuchila	Maggots in the wound	Bark	Apply the paste in the wounds with pus along with *Dodonaea angustifolia* L., f. Sapindaceae.
28.	*Terminalia arjuna* Roxb. Combretaceae/ Ghosh 295	Arjun	Bone fracture	Bark	Externally applied the lukewarm-paste as balm along with garlic bulb.
29.	*Trachyspermum ammi* L. Apiaceae/Ghosh 301	Jawan	Colic pain	Seed paste	Regular intake for 2 time in lukewarm water.

S.No.	Botanical name and Family/Collection No.	Local Name	Ailments	Parts Used and preparation	Mode of administration
30.	*Vernonia anthelmintica* L. Asteraceae/Ghosh 314	Somraj	Worm infestation	Fruit (20g)	Fed the paste of fruit with ginger and 1g camphor (*Cinnamomum camphora*, Lauraceae)
31.	*Vitex regundo* L. Verbenaceae/Ghosh 319	Buan/Begna	Foot and mouth disease	Leaf and twig	Decoction of plant used for washing the infected area.

Particularly out of 97; 75 tribal villages of Ajodhya hill of Purullia district is rich in ethno-veterinary knowledge owing to their close affinity with the surrounding plant cover. With the adoption of modern medicine the above mentioned ethnomedicines are gradually being abolished. Religions and cultural faith, poor economy etc. are the main cause for the utilization of these herbal drugs. Due to biotic interference some species are now gradually being endangered. Indigenous practices are the best way to keep the sheep well even in the contemporary age of allopathic treatment.

References

1. Jain SK and De JN, Some less known plant foods among the tribal of Purulia (West Bengal), Sci Cult. 1964, **30**, 285 – 286.

2. Ghosh A, Ethnoveterinary medicine from the tribal areas of Bankura and Medinipur district, West Bengal, IJTK, 2002, **1**, 93 – 95.

3. Ghosh AKG, Herbal veterinary medicine from the tribal areas of Midnapur and Bankura district, West Bengal, *J. Econ, Taxon, Bot,* 2003, **27**(3), 573 – 575.

4. Mukherjee A and Namhata D, Herbal veterinary medicine as practiced by the tribals of Bankura district, West Bengal, *J. Bengal Nat, Hist. Soc,* 1988, **7**, 69 – 71.

4.2 Ethnobotanical plants used in West Rarrh region both for domesticated animals and Human

Abstract

Present communication brings to knowledge about the indigenous methods of healing recorded from tribals and rural folks. A total of 60 plant species belonging to 34 families of angiosperms are executed by the inhabitants in the form of decoction, oil, paste, latex etc. either as a sole drug or in combination to treat various ailments. The dose/s, duration and method of administration are given along with botanical and local name, family, plant's part/form of recipe used. The folk herbal formulations however require further testing.

Keywords: Ethnobotany, West Rarrh, Decoction, Leukaemia

Introduction

Plants have been used as a source of medicine for living creatures since the time immemorial. Still maximum people in developing countries rely chiefly on indigenous medicines for primary healthcare. The habitants of this area have rich knowledge about plants utility and the literature survey shows that the region was almost untapped[1-3]. Here the tribals are mostly Santals, Lodha, Mahali, Munda, Bhumij, Oraons, Savars. Human involvement in depletion of the plant resources of the nature and rapid moderization of curative systems have gradually replaced the ancient but effective system of tribal medicine.

Materials and Methods

Ethnobotanical field survey was carried out during the years 2008 – 2010. Information on folk medicinal use of plants was obtained through oral interviews enduring local plant name, parts used, other ingredients added (if any), method of preparation and mode of administration for each species. Plant specimens were collected from the study area, authenticated and kept in the herbarium of the institute. The ethnobotanical information is given in Table 4.2

Results and Discussion

Locally available plants are used by the people as primary healthcare which still find place in their traditional therapy. However, isolation of active principles phytochemical and pharmacological investigations are desired to validate the claims of the traditional healers. The formulation of these effective phytomediciens should be encouraged for their sustainable uses. Statistically, information for treating a particular ailment from different information certainly reflect the accuracy and authenticity of the folk drugs employed. It is observed from the foregoing account that the tribals of West Rarrh use locally available plants alone or in various combinations to treat some common diseases of cattle. All the drugs are very

Table 4.1 : The etho-vetetinary medicinal information

A. For Domesticated Animals

Ailment	Plants used			How medicine is prepared	How used
	Local name	Botanical name with family	Parts used		
Expulsion of placenta	Bot	*Ficus benghalensis* L. (Moraceae)	Tender leaves	Banyan leaves (10) are grounded, mixed with 500 g bajra flour and made a paste with 1 litre water.	The formulation is given to cow after calving for 2 times.
	Bajra	*Pennisetum typhoides* (Burm. f) (Poaceae)	Flour		
Footsore	Akanda	*Calotropis gigantea* L. (Asclepiadaceae)	Latex	Collection of fresh latex	Smear the latex
Cut	Ayapan	*Eupatorium triplinerve* Vahl L. (Asteraceae)	Leaf	Fresh leaf juice	Check the bleeding instantly.
	Jam	*Syzygium cumini* L. (Myrtaceae)	Leaf	Fresh leaf juice.	Check the bleeding.
Prevention of sunstroke	Satamuli	*Asparagus racemosus* Willed (Liliaceae)	Leaves	200 g of leaves are soaked in water.	Applied on the head of cattle on alternate days.
Dropsy	Punarnaba	*Boerhavia diffusa* L. (Nyctaginaceae)	Plant	Plant decoction	Fed ½ cup once daily.
Shoulder wound and hoof's wound	Dhatri	*Woodfordia fruticosa* (L) Kurz. (Lythraceae)	Flower	Powder of the flower with coconut oil	Locally applied on the affected parts.
Ploughsh-are wound	Dhatura	*Datura metel* L. (Solanaceae)	Seed	Ripe seeds boiled in safflower oil	Apply in the leg consecutively for 4 days.

| Ailment | Plants used | | Parts used | How medicine is prepared | How used |
	Local name	Botanical name with family			
Broken horn	Nishinda	Vitex negundo L. (Verbenacae)	Leaf	Leaf dust	Smear the dust.
	Jam	Syzygium cumini L. (Myrtaceae)	Bark	Bark extract	Smear the extract as balm.
Blood dysentery	Jam	Do	Bark	Bark extract	Fed 1 cup once daily.
Poisonous food intake	Amrul	Oxalis corniculata L. (Oxalidaceae)	Leaf	Fresh leaf paste	Fed to the cattle 1 cup juice thrice daily.
	Talmuli	Curculigo orchioides Gaerth. (Amaryllidaceae)	Root	Root decoction along with fruit pulp of tamarind in 1:1 ratio.	Fed to the cattle 1 cup
	Satamuli	Asparagus racemosus Willd (Liliaceae)	Root Juice	Mixed it with banana root juice in 1:1	Fed 1 cup twice daily.
Hydrop-hobia	Kakmachi	Solanum nigrum L. (Solanaceae)	Fruit	Fresh extract of ripen fruits	Fed to the cattle 5 ml twice daily.
Indigestion	Kuchila	Strychnos nux-vomica L (Loganiaceae)	Seed	Dry seeds	Just after delivery to enhance digestion power a seed to fed to cattle once daily.
To achieve yellow yolk	Kuleykhara	Hygrophila spinosa T. Aders. (Acanthaceae)	Plant	Plant extract	1 ml fed once daily to the poultry birds.
	Methi	Trigonella foenum graceum L. (Fabaceae)	Leaf	Fresh leaves	Fed to poultry birds with food.

Ailment	Plants used			Parts used	How medicine is prepared	How used
	Local name	Botanical name with family				
To achieve pregnancy	Gokhur	*Tribulus terretris* L. (Zygophyllaceae)		Fruit	Fruit powder (100g) with germinated grams	Fed to the cows once daily at dawn.
Hoof erosion	Seuli	*Nyctanthes arbortristis* L. (Oleaceae)		Bark	Fresh extract	Smear the hooves.
	Chatim	*Alstonia scholaris* R. Br. (Apocynaceae)		Latex	Fresh latex	Smear the latex in the wound of hooves.
Flatulence	Talmuli	*Curculigo orchioides* Gaertn. (Amaryllidaceae)		Root	Decoction of roots	Fed to the cattle ½ cup twice daily.
Ague	Tulsi	*Ocimum sanctum* L. (Lamiaceae)		Dry leaves		Fumigate the cattle with the help of dry leaves.
Lactogenic	Anantamul	*Hemidesmus indicus* R. Br. (Asclepiadaceae)		Root	Boiled root (100 g) with fragmented rice (1 kg)	Fed to the cattle during lactation period.
Lactogenic	Methi	*Trigonella foenumgraecum* L. (Fabaceae)		Plant	Freshly collected plants 50 g for mother and 500 g for cattle	Fed to the cattle, mother once daily during lactation period.
Leukaemia	Bhringaraj	*Wedelia chinensis* L. (Asteraceae)		Leaf	Leaf juice mixed in milk	Fed to the patient and cattle 1 cup once daily.
Blood urea	Hakuch/ Latakasturi	*Psoralea corylifolia* L. (Fabaceae)		Leaf	Leaf juice with honey	Fed to the human and cattle once daily.

Ailment	Plants used		Parts used	How medicine is prepared	How used
	Local name	Botanical name with family			
Arthritis	Pinga	*Celastrus paniculatus* Willd. (Oleaceae)	Seeds	Oil from seeds	Smear the oil on painful organs both in cattle and man.
Pox	Methi	*Trigonella foenumgraceum* L. (Fabaceae)	Seed	Soaked water	Fed to the cattle 1 cup twice daily.
Sprain	Arani	*Clerodendron phlomides* L. f (Verbenaceae)	Leaf	Lukewarm paste	Apply on the affected parts twice daily.
Cold & cough	Basak	*Adhatoda zeylanica* Medic (Acanthaceae)	Leaf	Fumigation of dry leaves	Fumigate the cattle twice daily.
Bone fracture	Tentul	*Tamarindus indica* L. (Caesalpiniaceae)	Seeds	Soaked 200 g of seeds in water for 4 hrs and then grinds into a paste and boiled.	Smear the paste over fractured leg covered with coconut husk.
	Ghritakumari	*Aloe vera* (L.) Burm. f. (Liliaceae)	Leaf	Leaf extract	Smear as balm on broken part.
	Kathila	*Sterculia urens* Roxb. (Sterculiaceae)	Bark, Latex	Bark paste	Smear as balm.
Eye ailment	Shetdrone	*Leucas aspera* Spr (Lamiaceae)	Leaf	Made a paste in goats milk from 100 g leaves with 10 g black pepper.	Paste is applied twice daily.
Foot & mouth disease	Sondal	*Cassia fistula* L. (Caesalpiniaceae)	Bark	Bark decotion	Wash the wounds with decoction.

Ailment	Plants used				How used
	Local name	Botanical name with family	Parts used	How medicine is prepared	
	Lajjabati	*Mimosa pudica* L. (Mimosaceae)	Root	Fresh root extract	Smear the extract.
Tympany (bloat)	Arjun	*Terminalia arjuna* Roxb. (Combretaceae)	Bark	Made a decoction dissolving 100 g powder in 250 ml of lukewarm water.	Fed the cattle 10 ml twice daily.
Diarrhoea	Bans	*Bambusa arundinacea* (Retz.) (Poaceae)	Leaf	Made a paste from 10 leaves each.	Fed to the cattle once a day to check the worm loaded diarrhoea.
	Kanthal	*Artocarpus heterophyllus* Lamk. (Moraceae)	Leaf		
	Khayer	*Acacia catechu* L. (Minosaceae)	Bark	250 g bark powder of each boils in a litre of water.	Fed to the cattle the filtrate formulation thrice daily.
	Palash	*Butea monosperma* (Lom)(Fabaceae)	Bark		
	Piyal	*Buchanania lanzan* sprong (Anacardiaceae)	Bark		
	Kunch	*Abrus precatorius* L. (Fabaceae)	Seeds	Seed poweder (10 g)	Fed to the cattle once daily.
Wound	Musambi	*Citrus sinensis* Osbeck (Rutaceae)	Leaf	Made a paste from 15 leaves and applied on wounds	Use the paste thrice daily on the wounds of cattle.

| Ailment | Plants used | | | How medicine is prepared | How used |
	Local name	Botanical name with family	Parts used		
	Satin wood	*Chloroxylon swietenia* DC (Meliaceae)	Root	Paste of roots give as bolus (pill)	Fed twice a day to kill the maggols.
	Ayapan	*Eupatorium triplinerve* Vahl (Asteraceae)	Leaf	Leaf extract	Smear the juice on the wound.
	Iswarmul	*Aristolochia indica* L. (Aristolochiaceae)	Leaf	Leaf extract	Smear the juice on the wound.
Uterine prolapse	Shet Lajyabati	*Mimosa pudica* L. (Mimosaceae)	Leaf	500 g fresh leaves is fed to buffaloes once daily.	
Worm infestation	Neem	*Azadirachta indica* A. Joss. (Meliaceae)	Leaf	100 g each of the crushed leaves soaks in 400 ml water for an hour	The formulation is shake well before use and given to cattle once a day for 2-3 days.
	Dalim	*Punica granatum* L. (Punicaceae)			
Parasitic infestation	Palash	*Butea monosperma* (Lamk) (Fabaceae)	Seeds	Made a paste	Fed to cattle once a day for 3 days
	Boch	*Acorus calamus* L. (Araceae)	Rhizome	Leachate	Smear the water on demestic birds and cattle.
Ticks	Kamala lebu	*Citrus reticulata* Blanco (Rutaceae)	Leaf	Collection of fresh juice in an earthen pot for 2 days and then applied to the affected cattle.	After 2 days the cattle is bathed thoroughly.

Ailment	Plants used				How used
	Local name	Botanical name with family	Parts used	How medicine is prepared	
	Parashi	*Cleistanthus collinus* (Roxb.) (Euphorbiaceae)	Leaf	Leaf extract	Smear once daily in domestic birds and cattle
	Bhringaraj	*Wedelia chinensis* L. (Asteraceae)	Leaf juice	Mixed it with tobacco leaf leachate	Smear once daily with a pinch of camhor on birds.
	Ulat – chandal	*Gloriosa superba* L. (Liliaceae)	Root	Fresh root extract	Smear on bird and cattle.
Filaria	Aparajita	*Clitoria ternatea* L. (Fabaceae)	Root	Paste of root	Smear the paste
Cour wound	Shet akanda	*Colotropis gigantea* (L.) R. Br. (Asclepiadaceae)	Latex	Fresh latex	Smear the latex on wound.
Conjunc-tivitis	Apang	*Achyranthes aspera* L. (Amaranthaceae)	Root	Root leachate water with sour curd	Externally applied to the affected eyes.
Fertility/ promoting conception	Kontikeri	*Solanum surattense* Burm. f. (Solanaceae)	Leaf	Leaf and plant juice	Fed 2 teaspoonful once daily.
Jaundice	Kalmegh	*Andrographis paniculata* Burm. f. (Acanthaceae)	Leaf	Leaf juice	Fed 2 teaspoonful juice once daily.

| Ailment | Plants used | | Parts used | How medicine is prepared | How used |
	Local name	Botanical name with family			
Obesity	Kurtikalai	*Dolichos biforus* L. (Fabaceae)	Seed	Fried in mustard oil	Rubbed the oil before bed time.
Urinary stone	Do	Do	Seed	Seed leachate water	Drunk 1 cup once daily
Intoxication	Kuchila	*Strychnos nux-vomica* L. (Loganiaceae)	Seed	Chyme prepared from the seeds boiled in milk	Fed 1 chyme (25 g) once daily.
Memory retention	Chita	*Plumbago zeylanica* L. (Plumbaginaceae)	Root	Root powder mixed in ghee with slight honey	Fed the formulation once daily.
Acne in tongue	Telakucha	*Coccinia grandis* L. (Cucurbitaceae)	Fruit	Masticated green fruit	Follow the practice
Madness	Boch	*Acorus calamus* L. (Araceae)	Root	Root extract mixed in honey	Fed to the patient 5 ml twice daily
Easy parturition	Ulatchandal	*Gloriosa superba* L. (Liliaceae)	Root	Paste smear on palm and foot	Helps placental retention with parturition
Jaundice	Bhui – amla	*Phyllanthus niruri* auet. (Euphorbiaceae)	Plant	plant juice with sour curd.	Fed ½ cup once daily.

C. For Plant

Ailment	Plants used			How medicine is prepared	How used
	Local name	Botanical name with family	Parts used		
Flower shedding in vegetables pulses, oil seeds and pine apple.	Hing	*Ferula asafoetida* L. (Apiaceae)	Resin of the root	50 g of hing is mixed with 500 g of flour and put in a cloth bag	The bag is put in the irrigation canal at the time of blooming which compensate the sulphur deficiency.
	Bajra	*Pennisetum typhoides* (Burm, f) (Poaceae)	Flour		
Aphids in cotton	*Teshiro monsha*	*Euphorbia antiquorum* L. (Euphorbiaceae)	Latex	Made a solution of 15 litres water mixed with 100 g latex	Sprays the solution twice a month.
Purification of drinking water	Anantamul	*Hemidesmus indicus* (L.) R. Br. (Asclepiadaceae)	Leaf dust	Mixed 1 g/litre	As a preventive measure against enteric & dysentery.
	Boch	*Acorus calamus* L. (Araceae)	Rhizome	Mixed 1g/litre	Do
	Bel	*Aegle Marmelos* (L.) Corr. (Rutaceae)	Fruit pulp	Beverage from the pulp	Drunk the beverage to resist cholera.

effective, and cheaply available in West Rarrh as alternative to allopathic medicines. The treatment shows no adverse side effect.

References

1. Jain S K, Observations on ethnobotany of tribals of Madhya Pradesh, Vanyajati, 1963, II, 177 – 183.

2. Cox P A, In : Ethnobotany and the Search of New Drugs, by DJ Chadwick and J Marsh (Eds), John Wiley & Sons, England 1994, pp. 25 – 36.

3. Ghosh A, Ethnomedicinal plants used in West Rarrh region of W.Bengal, Natural Product Radiance, 2008, 7(5), 461 – 465.

5. Glossary of Ecological Terms

1. **Afforestation :** It is a process in which a large scale tree planting is carried out in open land and barren hills.

2. **Biodegradable :** Certain organic or inorganic substances could be easily destroyed by micro-organisms through decomposition process.

3. **Biomass :** Total weight of all the organisms in a particular habitat.

4. **Biosphere :** The envelope containing all the living things on earth.

5. **BOD :** Biological Oxygen demand. It is the amount of O_2 required by micro-organisms in the water to carry out the decomposition process aerobically used as a measure in determining the contamination status of the water.

6. **Biotic Stability :** Ability of the population to live in an equilibrium or stable condition with an environment even after disturbance.

7. **Calorimeter :** An instrument used to measure the amount of energy in a given substance.

8. **Canopy :** The upper portion of the tree viewed from the sky.

9. **Climax Vegetation :** The last stage but the most stable condition within an ecological succession of communities.

10. **Decomposers :** Micro-organisms which lead the life by breaking down the dead organic matters in order to obtain energy.

11. **Deforestation** : The process by which trees are being out.

12. **Desertification** : Creation of desert like situation artificially by human activity.

13. **Ecology** : The study of the relationship between organisms and their environment.

14. **Ecosystem** : A system in ecology where organisms interact with each other for their survival.

15. **Ecotone** : Zone of transition between 2 vegetational regions.

16. **Energy Crops** : Crops grown to obtain energy from them (*e.g.* sugarcane provides ethanol, a potential substitute for conventional fuel such as petrol).

17. **Energy Flow** : Trapping of solar energy by photosynthetic organisms and its transformation into chemical energy and passes through various trophic levels.

18. **Environment** : Sum of all external forces or influences that affect an organism.

19. **Euphotic Zone** : A layer of the water body especially the surface where organisms can photosynthesize with the help of light.

20. **Eutrophication** : The process occurring when the level of nutrients (organic) in a body of water increases markedly leading to an increase in the number of aquatic organisms.

21. **Field-Water Capacity** : The maximum amount of capillary water that a particular soil is able to hold.

22. **Forestry** : Science which concentrate on how to produce forest goods.

23. **Gene Pool** : Collection of genetically similar or dissimilar organisms.

24. **Grazing Food Chain** : The food chain in which animals eat the primary producers.

25. **Green House Effect** : Ability of the atmosphere to act like a green house where heat is retained by the glass wall. In atmosphere certain gases trap and retain the heat.

26. **Green Manure** : Nutrients produced from the green plants to use as fertilizers.

27. **Gross Primary Production** : The rate at which solar energy is converted into chemical energy per unit of earth surface per unit time.

28. **Habitat.**: Natural environment of plant or animal.

29. **Humus** : Organic matters formed in the forest soil due to decomposition of fallen leaves.

30. **Hydrosere** : Succession which starts from the aquatic system.

31. **Ice Cap Analysis** : Measurement of CO_2 concentration in the ancient snow on ice caps of glaciers.

32. **Laterite** : Reddish infertile tropical soil because of intense leaching out of silica; only Fe and Al rich clay is left, remains fertile only when a considerable amount organic matter is added continuously.

33. **Lentic Water** : Standing water bodies such as water of ponds or pool etc.

34. **Lianes** : Twining vines with woody stems.

35. **Litter** : A layer of undecomposed plant parts on the surface of the ground.

36. **Lotic Water** : Running water bodies of river etc.

37. **Monocroping** : Practice of cultivating only one kind of crop plant for many seasons.

38. **Monoculture** : Agricultural proctice of producing or growing a single crop or plant species over a wide area and for a large number of sonsecutive years.

39. **Mulching** : Practice of covering surface soil for maize stalks, cotton stalks, tobbacco stalks, potato tops etc.

40. **Natality** : Death rate of a population.

41. **Net Primary Productivity** : The gross primary production minus, the amount of chemical energy used in respiration.

42. **Nuclear Waste** : Used radioactive substances of nuclear power stations.

43. **Oil Slick** : The floating layer of oil on the surface of water bodies.

44. **Ozone** : The thin layer which forms as a part of the upper atmosphere which filters out ultra-violet light from the sun.

45. **Polyploidy** : A process in which chromosome number doubles due to failed meiosis which results in the formation of new species.

46. **Population** : Group of related individuals capable of inter breeding.

47. **Primary Succession** : A succession beginning with bare or uninhabited area.

48. **Productivity** : Rate of dry-matter production by photosynthesis in an ecosystem.

49. **Pugmark** : Foot print of wild animals. During survey by measuring the pugmark the population of tiger, lion etc. can be detected.

50. **Pyrolysis** : Subjecting biomass to 400°C to extract energy from it.

51. **Quadrat** : A frame of any shape, which when placed over vegetation defines a unit sample area within which the plants may be counted.

52. **Radiation** : Stream of particles or rays which are emitted by certain naturally occurring elements like U^{235}, radium 226 etc.

53. **Remote Sensing** : Technique of taking photograph of landscape from the airborne or space borne platform.

54. **Savanna :** A tropical grassland consisting of tall, coarse grasses and scattered small trees.

55. **Secondary Succession** : A succession that begins with the disturbed remains of a previous vegetation.

56. **Soil Profile :** The structure and composition of a soil as seen in the side of a pit, trench etc., made up of layers of horizons produced by leaching and deposition.

57. **Steady State :** State in which the inflow of energy and materials in an ecosystem is just sufficient to maintain biomass at a relatively constant level.

6. Glossary of Medical Terms

Abscess : A localized collection of pus in any part of the body. The result of disintegration of displacement of tissue.

Abortifacient : Anything used to cause or induce an abortion.

Ague : A popular name for malarial fever.

Alexeteric : Protective against infection, venom and poison.

Alexephermic : Antidotal.

Alternative : A drug which corrects disorder process of nutrition and restores the normal function of an organ or of the system.

Amenorrhoea : Suppression of menses not due to natural causes.

Anaemia : Condition in which there is a reduction in the number of circulating red blood per cu. mm, the amount of haemoglobin per 100 ml of blood.

Anodyne : A drug that relieves pains : an analgesic.

Anthelmintic : An agent that destroys parasitic intestinal worms.

Antithydrotic : A drug which checks sweating.

Antiperiodic : A drug which controls periodic attacks of disease; an anti-malarial drug.

Antispasmidic : A drug which prevents or cures colic, convulsions or spasmodic disorders.

Antipyretic : Reducing fever. An agent that reduces fever.

Antidote : A substance that neutralizes poisons or their effect.

Antisyphilitic : Curative or relieves syphilis.

Antiseptic : An agent capable of producing antiseposis.

Anticholinergic : Impending the impulses of cholinergic neurotransmitters. An agent that blocks parasympathetic nerve impulses.

Anthrax : Acute, infectious disease caused by *Bacillus anthracis*, usually attacking cattle, sheep, horses and goats. Man contracts it from contract with animal hairs, hides or waste.

Aperient : A laxative or mild purgative.

Aphrodisiac : A drug which promotes sexual desire.

Apoplexy : Sudden loss of consciousness followed by paralysis caused by haemorrhage in the brain.

Aromatic : A drug having an agreeable odor.

Ascarides : A collection of fluid in the abdomen, abdominal dropsy.

Asthma : A disease of the bronchial tubes causing recurrent attacks of breathlessness and coughing.

Astringent : To bind fast, drawing together, constricting, binding. An agent that has a constricting or binding effect; *i.e.,* one that checks haemorrhage or secretions by coagulation of proteins on a cell surface.

Benzaldehyde : A pharmaceutical flavouring agent derived from oil of better almond.

Billiousness : A symptom of a disordered condition of the liver causing constipation, headache, loss of appetite and vomiting of bile.

Black water fever : Billious remittent fever, a complication of malaria.

Belennorrhoea : Excessive mucous discharge, particularly from the urethra or vaginal gonorrhoea.

Blister : An accumulation of the fluid under the upper layer of the skin; a substance applied to the skin for raising a blister.

Boil : An infectious festering sore which ultimately may develop into an ulcer.

Bright's disease : An acute or chronic disease of kidneys.

Bronchitis : An inflammation of mucous membrane of the bronchial tubes or air passage; feverish cold with cough and sore chest.

Bruises : An injury with diffuse effusion into subcutaneous tissue and in which skin is discoloured but not broken.

Cardiac : Pertaining to heart.

Cardiotonic : Increasing tonicity of the heart. Various drugs, including digitalis, are cardiotonic.

Carminative : A drug which relieves flatulence or the feeling of over-fulness of the stomach.

Catarrh : Term formerly applied to inflammation of mucous membranes, especially of head and throat.

Cathartic : An active purgative producing bowel movements.

Cerebral : Pertaining to cerebrum, the brain.

Cholangogue : An agent that increases the flow of bile into the intestine.

Chorea : A nervous condition marked by involuntary muscular twitching of the limbs or facial muscles.

Colic : Spasm in any bollow or tubular soft organ accompanied by pain.

Cutaneous : Pertaining to skin.

Cystitis : Inflammation of bladder.

Dandruff : Normal exfoliation of the epidermis of the scalp in the form of dry white scales.

Demulcent : Stroking softly. An agent that will smooth the part of soften the skin to which applied.

Depressant : An agent that reduces functional activity.

Delirium : A state of mental confusion and excitement characterized by disorientation.

Depurative : Having cleaning properties; removing waste material from the body. Removal of waste material.

Deodorant : Perfumoral. An agent that masks or absorbs foul odours.

Diaphoretic : An agent that increase perspiration, such as camphor, opium or pilocarpine. Heat may also be included as such an agent. An agent that induces copious secretion of sweat.

Diarrhoea : Frequent passage of unformed watery bowel movement. It is a frequent symptom of gastrointestinal disturbance.

Diathesis : Constitutional predisposition to a certain disease, condition or group of diseases. Can be allergic, haemorrhagic or rheumatic disease.

Diabetes : A general term for diseases characterized by excessive urination. Usually, refers to Diabetes mellitus.

Dropsy : Disease causing a watery fluid to collect in some cavity of the body.

Diuretic : An agent that increases the secretion of urine.

Dyspepsia : Imperfect or painful digestion. Not a disease in itself but symptomatic of other diseases or disorders, indigestion.

Dysentery : An infectious disease, characterized by acute diarrhoea accompanied by gripping pains, the stools being chiefly of blood and mucous.

Dysuria : Painful and difficult urination.

Dysmenorrhoea : Pain in association with menstruation. One of the most frequent gynaecologic disorders.

Eczema : Itching skin disease.

Elephantiasis : A disease of the skin and subcutaneous tissues causing hypertrophy of the affected parts. It may attack any part of the body but chiefly attacks the legs.

Emetic : An agent that produces vomiting.

Emollient : An agent that will soften and soothe the part when applied locally.

Emmenagogue : A substance that promotes or assists the flow of menstrual fluid.

Epilepsy : Nervous disease causing a person to fall unconscious (often with violent involuntary movement).

Expectorant : An agent that facilitates the removal of the secretion of the bronchopulmonary mucous membranes. Classified as sedative or stimulating.

Febrifuge : An agent or drug used for reducing fever.

Febrile : Pertaining to fever, feverish.

Flatulence : Excessive gas in the stomach and intestines.

Freckles : Coloured spots, generally yellowish or brown, on the exposed parts on the skin.

Fructose : Levulose, fruit sugar.

Galactagogue : An agent that promotes the secretion and flow of milk; lactagogue.

Glect : A mucous discharge from the urethra in chronic gonorrhoea.

Gonorrhoea : A specific contagious, catarrhal inflammation of the genital mucous membranes of either sex.

Gravel : The development in the kidneys and urinary tract of tiny stone-like collection of uric acid, calcium oxalate or phosphates.

Granulation : Formation of granules or slate or condition of being granular.

Griping : A sharp policy pain in the bowels due to presence of some irritating substance.

Haematuria : Passing of blood in the urine.

Haemorrhage : Bleeding, especially profuse, from any part of the body.

Haemorrhoids : Piles; a diseased condition of the blood vessels causing painful swellings in the region of the anus.

Hemicrania : Headache on only one side of the head; migraine.

Hepatitis : Inflammation of the liver.

Herpes : Creeping skin diseases.

Hernia : The protrusion or projection of an organ or a part of an organ through the wall of the cavity that normally contains it.

Hiccough Hiccup : Spasmodic periodic closure of the glottis following spasmodic lowering the diaphragm, causing a short sharp, inspiratory cough.

Hydrophobia : Morbid fear of water. Common name of rabies resulting from bite of rabid animal.

Hysteria : A disease in which the patient who is physically healthy, suffers from imaginary diseases and has lost control over acts and feelings.

Hypnotic : Pertaining to sleep or hypnosis. An agent that induces sleep or that dulls the senses, such as chloral hydrate.

Induration : Area of hardened tissues.

Insomnia : An excessive amount of fibrin in the blood.

Intermittent : Suspending activity and intervals. Coming and going.

Ipacacauanha : It is a plant grown in Brazil. The dried root of this plant is know as Ipacacauanha. It is the source of emetine q.v.

Itch : An infectious disease of skin without specific lesions and marked by excessive itching.

Jaundice : A disease condition of the liver in which there is yellowish colouring of the tissues and urine with bile.

Laryngitis : Inflammation of larynx.

Leprosy : A chronic wasting disease caused by a germ which generally, results in mutilations and deformities.

Leucoderma : A skin disease marked by skin losing its pigment wholly or partially.

Leucorrhoea : A vaginal discharge of a white fluid containing mucous or pus cells.

Lithiasis : The formation of calculi or stone in any part of the body.

Lithotriptic : A drug having the property of crushing a calculus in or stone present in the urinary system.

Lithositic : Having the property of crushing a calculus in the bladder or urethra.

Lumbago : Rheumatism of the small of the back causing acute pain and stiffness.

Mania : Mental disorder characterized by excessive excitement.

Melancholia : A disorder of the mind marked by depression of spirits, mental sluggishness and apathy to ones surroundings.

Menorrhagia : Abnormally excessive menstruation.

Micturition : Urination.

Migraine : A nervous disorder marked by periodic attacks of headache; hemicrania.

Mucilaginous : Resembling mucilage, slimy; sticky.

Mumps : An infectious disease marked by inflammation of glands near the ear.

Mydriatic : A drug which dilates the pupil.

Narcotic : A drug which induces deep sleep or insensibility to pain.

Nausea : A feeling of sickness, inclination to vomit.

Nephritis : Inflammation of the kidneys.

Neuralgia : Severe sharp pain along the course of nerves.

Opthalmia : Inflammation of the nerves, conjunctivitis.

Otitis : Inflammation of the ear.

Otorrhoea : A purulent discharge from the ear.

Palsy : Temporary or permanent loss of sensation or loss of ability to move or to control movement.

Paralysis : Loss of motion in any fraction of any part of the body.

Paroxysm : A sudden, periodic attack or recurrence of symptoms of a disease. Sudden spasm or convulsion of any kind. Sudden emotional state, as of fear, grief or joy.

Pectoral : Pertaining to the chest; cough remedy; expectorant.

Photophobia : Unusual intolerence of light.

Phithisis : Affected with pulmonary tuberculosis.

Piles : An inflammed condition of veins in the rectal region; haemorrhoids.

Pleurisy : Inflammation of the membrane enclosing the lungs.

Pneumonia : Inflammation of the lungs.

Polyploidy : Condition in which the chromosome number is two or more times the normal haploid number found in gametes.

Poultice : A hot, moist mass of linseed, mustard or soap and oil between two pieces of muslin applied to the skin to relieve conjection or pain, to stimulate absorption of inflammatory products, and to act as a counter irritant.

Prolapse : A falling down of an organ from its normal position, especially its appearance at an opening.

Pruritus : Severe itching. May be a symptom of a disease process such as allergic response or be due to emotional factors.

Pulmonary : Pertaining to lungs.

Pungent : Sharp smell or taste.

Pyorrhoea : A purulent discharge from the gums.

Pyrosis : A burning sensation in the epigastric and sternal region with raising of acid liquid from stomach.

Rectum : Lower part of large intestine.

Refrigerant : An agent which relieves feverishness or produces a feeling of coolness.

Refringent : Refractive.

Rheumatism : A term used for pains in the muscles, joints and certain tissues; the disease takes various forms.

Resolvent : Promoting disappearance of inflammation.

Rubefacient : A mild counter-irritant; a drug that causes tingling reddening of the skin.

Sciatica : A neuralgic pain at the back of the thigh caused by the inflammation of the sciatica nerve.

Scorbutic : Suffering from scurvy.

Scrofula : A disease chiefly of the young, marked by want of resisting power making the patient susceptible to tuberculosis especially of the glands, bones and joints, eczematous eruption, ulceration, glandular swelling, etc.

Sedative : A drug which has a claiming or quieting effect on the patient and which reduces nervous excitement.

Sialagogue : A drug which promotes the secretion of saliva.

Soporific : A drug that induces sleep.

Spasmophile : A tendency to tetany and convulsion; almost always association with rickets.

Spasmodic : A convulsion. Concerning spasms.

Stomatitis : Inflammation of the mucous membrance of the mouth.

Stomachic : A drug which improves digestion and appetite.

Strangury : Painful and drop by drop discharge of urine.

Styptic : An agent which checks bleeding.

Sudroific : An agent that promotes perspiration; a diaphoretic.

Suppuration : The process of pus formation.

Syphilis : A serious chronic venereal disease.

Taenia : Tapeworms.

Tetanus : An infectious disease, marked by painful contractions in the muscles.

Tinea : A group of parasitic skin diseases, ringworm.

Tonsillitis : Inflammation of tonsil.

Trachoma : A contagious granular inflammation of the conjective.

Tuberculosis : A disease caused by bacillus; it may affect any part of organ of the body.

Ulcer : An open sore on the skin or on any mucous membrane.

Vermifuge : A drug which expels intestinal worms.

Vertego : Dizziness; giddiness.

Vesicant : A drug or agent that produces blisters.

Vulnerary : A drug which promotes healing of wounds.

Wart : A hypertrophy or growth on the skin.

Whooping cough : An acute infectious disease marked by recurring peculiar spasmodic attacks of coughing, each attack ending with a deep noisy intake of breath.

7. List of Publications

1. Agronomy and Crop Science 162, 342 – 346 (1989). 1989 Paul Parey Scientific Publishers Berlin and Hamburg ISSN 0931 – 2250. Monocarpic Senescence in Relation to Yield of *Sesamum indicum* During Source-Sink Alternation, A.K. Biswas and A.K. Ghosh.

2. *Indian J. Plant Physiol.*, Vol, XXXII, No. 2, pp. 172 – 174 (June, 1989). Monocarpic Senescence of *Arachis hypogaea*: Nutrient Withdrawal Vs. Migration of Senescence Signal, A.K. Ghosh and A.K. Biswas.

3. Indian Journal of Experimental Biology, Vol. 28 May 1990, pp. 492 – 493. Whole plant senescence in *Ophyioglossum vulgatum:* Influence of sporangiferous spike, kinetin and abscisic acid. Arun Kumar Biswas, Ashis Kumar Ghosh, Swapan Kumar Mandal and Udayan Sarkar.

4. *Indian J. Plant Physiol.*, vol. XXXIV No. 1, pp. 25–29 (March, 1991). Source – Sink relationship during Ageing and Senescence of *Solanum tuberosum* L. Ashis Kumar Ghosh and Arun Kumar Biswas.

5. Agri. Biol. Res. 7(2) 132 – 138 (1991). Source – Sink Relationship during Monocarpic Senescence of *Raphanus sativus.* A.K. Ghosh and A. Biswas.

6. An article the Botanica, Vol, 42, pp. 20–23 (1992) Magazine of Delhi Univ. Botanical Society : Inhibitory Effect of Fishes Exposed to some Cultivated, Wild plant from the Bangiya West Rardh (Bankura district). A.K. Ghosh.

7. Indian Journal of Forestry, Vol. 16(3) 201–203. 1993. Intact leaf senescence in some polycarpic forest trees in relation

to their reproductive development. Ashis Kumar Ghosh and Arun Kumar Biswas.

8. *Geobios New Reports* 12 : 94–96, 1993. Herbal folk medicines of Bankura district, West Bengal. D. Namhata and A. Ghosh.

9. *Geobios New Reports* 12 : 96–99, 1993. Effect of polluted air of Durgapur on the vegetation. Ashis Kumar Ghosh.

10. *Revista Biol* (Lisboa) 15 : 63–68, 1994. Monocarpic senescence in *Amaranthus tricolor L. : Dual action of kinetin.* A.K. Ghosh and A.K. Biswas.

11. Indian Journal of Experimental Biology. Vol. 32. November 1994, pp. 807–811. Correlative senescence in *Targets patula and Chrysanthemum coronarium* during reproductive development. Ashis Kumar Ghosh and Arun Kumar Biswas.

12. J. Agronomy and Crop Science 175 – 202 (1995). (c) 1995 Blakwell Wissenshafts Verlag, Berlin ISSN 0931 – 2250. Regulation of Correlative Senescence in *Arachis hypogaea* L. by Source-Sink. Alteration through Physical and Hormonal Means. A.K. Ghosh and A.K. Biswas.

13. J. Econ Taxon Bot additional Series, 12. Scientific Publishers, Jodhpur (India), 1996, pp. 318 – 320. Ethnomedicine in Bankura district, West Bengal. Ashis Ghosh, Sarathi Maity and Malaty Maity.

14. J. Econ. Taxon. Bot. Vol. 23 No. 2 (1999), pp. 557 – 560. Herbal Veterinary medicine from the Tribal Area of Bankura District, West Bengal, A.K. Ghosh.

15. Pak J. Sci. Ind Res 2002 45 (3) 212. Mechanism of Monocarpic Senescence of *Trichosanthes dioica* Roxb (Cucurbitaceae). Ashis Ghosh.

16. Ghosh Ashis Kumar (2002). "Herbal Veterinary Medicine from the Tribal Areas of Bankura and Midnapur District." In : Series Recent Progress in Medicinal Plants. Vol. 1, Ethnomedicine and Pharmacognosy (Eds : V.K. Singh, J.N. Govil and Gurdip Singh). Sci Tech Pub., USA 233 – 237 pp.

17. Indian Journal of Traditional Knowledge, Vol. I (i) October, 2002, pp. 93 – 95. Ethno-veterinary medicines from the tribal areas of Bankura and Medinipur districts, West Bengal, Ashis Ghosh.

18. J. Econ. Taxon. Bot. Vol. 27 No. 3 (2003). Scientific Publishers (India), pp. 573 – 575. Herbal veterinary medicine from the tribal areas of Midnapur and Bankura district, West Bengal, Ashis Kumar Ghosh.

19. J. Econ. Taxon. Bot. Vol. 27 No. 4 (2003). Scientific Publishers (India), pp. 825 – 826. Traditional vegetable dyes from central West Bengal, Ashis Ghosh.

20. Indian Journal of Traditional knowledge, Vol. 2 (4) October, 2003, pp. 393 – 396. Herbal folk remedies of Bankura and Medinipur districts, West Bengal, Ashis Ghosh.

21. Natural Product Radiance Vol 3 (2), p. 91 March-April 2004. Plant and Clay dyes used by Weavers and Potters in West Bengal, Ashis Ghosh.

22. Natural Product Radiance Vol. 3(3), p. 170 May-June 2004. Apiphilic plants in agro-forestry. Ashis Ghosh.

23. Natural Product Radiance Vol. 3 (3), p. 173 May-June 2004. Natural biocides and biofertilizers. Ashis Ghosh.

24. Natural Product Radiance Vol. 3 (6), p. 426 Nov-Dec. 2004. Home made baby food. Ashis Ghosh and Jhuma Karan.

25. Pak J. Sci Ind Res. 2005 48(1) 55 – 56. Mechanism of *Monocarpic Senescence of Momordica dioica*. Source – Sink Regulation by Reproductive Organs. Ashis Ghosh.

26. Non-Wood News, No. 12, p. 30 March, 2005, Rome, Italy, Plant and Clay Dyes. Ashis Ghosh.

27. Ageing and Society; The Indian Journal of Gerontology, Vol. 15(3&4), pp. 75 – 81, 2005. How one can survive for a long time in the earth. Ashis Ghosh.

28. Ghosh Ashis (2006). Identification of veterinary Medicinal plants through participatory approach. In : Advances in Medicinal Plants (Eds. N.D. Prajapati, T. Prajapati and S. Jaipura) Vol. 2 pp. 97 – 105. Asian Medicinal Plant and Health Care Trust, Jodhpur, Rajasthan.

29. Natural Product Radiance Vol. 5 (4) p. 260. July-Aug. 2006. Healthy Pan-drink. Ashis Ghosh.

30. J. Econ. Taxon. Bot. Vol. 30 (Suppl.), pp. 233 – 238 (2006). Medicinal plants used for treatment of diabetes by the tribals of Bankura, Purulia and Medinipur of W. Bengal. Ashis Ghosh.

31. Natural Product Radiance Vol. 7(5), 461 – 465, Sept – Oct 2008. Ethnomedicinal plants used in West Rarrh region of West Bengal. Ashis Ghosh.

32. Geobios 35 : 286 – 288, 2008. Source – Sink relationship during monocarpic senescence of *Pachyrhizus angulatus* (Papilionaceae) and *Papaver somniferum* (Papaveraceae). Ashis Ghosh.

33. J. Phytol. Res. 21(1) : 153, 2008. Tailoring of haulms serve as a natural remedy against Late Blight of Potato and natural resource for biofertilizer-An Observation. Ashis Ghosh.

34. J. Phytol. Res. 21(1) : 155, 2008. Ancient indigenous abolishing paddy varieties of Bengal. An Observation Ashis Ghosh.

35. Natural Product Radiance Vol. 8(3), p 213 May – June 2009. Traditional food preparation from Urd dal in Purba Medinipur, Ashis Ghosh.

36. Natural Product Radiance Vol. 8(5), p. 477 Sept – Oct. 2009. How to get honey in off season from detached plants? Ashis Kumar Ghosh.

37. Natural Product Radiance Vol. 8(5), p. 477 Sept. – Oct. 2009. Allelopathic susceptibility of some climbers. Ashis Kumar Ghosh.

38. J. Econ. Taxon. Bot. Vol. 33(2), pp. 321 – 323 (2009). Comparative study of monocarpic senescence of a monocot (*Commelina benghalensis* L.) and a dicot (*Oxalis corniculata* L.) : Migration of senescence signal Vs. nutrient drainage. Ashis Kumar Ghosh.

39. J. Econ. Taxon. Bot. Vol. 33(2), pp. 333 – 334 (2009). A note on quick regeneration of stem cuttings of bamboos as influenced by planting posture and IAA. Ashis Kumar Ghosh.

40. J. Econ. Taxon. Bot. Vol. 34(1), pp. 69 – 71 (2010). Indigenous knowledge on various fruit pulp processing of Suri in Birbhum district of W. Bengal. Ashis Ghosh.

41. J. Econ. Taxon. Bot. Vol. 34 (1), pp. 87 – 90 (2010). Natural Dye making process along with its Dye yielding plants of West Rarrh. Ashis Ghosh.

42. J. Econ. Taxon. Bot. Vol. 34(1), pp. 91 – 93 (2010). A Note on Natural antioxidants : The precursor of antiageing. Ashis Ghosh.

43. J. Econ. Taxon. Bot. Vol. 34 (1), pp. 94 – 95 (2010). Study of Ecoflame in Black variety of *Capsicum annuum* L. Ashis Ghosh.

44. J. Econ. Taxon. Bot. Vol. 34(3), pp. 635 – 638 (2010). Herbal remedies for alopecia used by the Tribal of West Bengal. Ashis Ghosh.

45. J. Econ. Taxon. Bot. Vol. 34(4), pp. 818 – 820 (2010). Traditional knowledge on auspicious plants for Ancestral abode in Ecological perspective. Ashis Ghosh.

46. J. Econ. Taxon. Bot. Vol. 35(2), pp. 424 – 428 (2011). Ethnoveterinary Medicine as Practice by the Tribals of West Rarrh. Ashis Kumar Ghosh.

Books

1. Regulation of Senescence in Various Plant by A.K. Biswas and A.K. Ghosh, Emkay Publication, Delhi.

2. Paribesh O Udvid – Ashis Ghosh.

3. Prani Rahasya – Ashis Ghosh.

4. Ethnobiology : Therapeutics and Natural Resources. Ashis Ghosh, Daya Publishing House, New Delhi.

5. Ethnomedicine for Human and Veterinary Development. Ashis Ghosh, Daya Publishing House, New Delhi.

6. Agronomy : Facts and Approaches. Dr. A.K. Ghosh and Dr. P. Das, Daya Publishing House, New Delhi.

7. Traditional Knowledge of Household products Dr. A.K. Ghosh, Daya Publishing House, New Delhi.

8. Paribesh O Parthiba Bastu.

Index

Index of Latin Names

INDEX OF DISEASE